Melvil Dewey

Library School Rules

1. Card Catalog Rules; 2. Accession Book Rules; 3. Shelf List Rules

Melvil Dewey

Library School Rules
1. Card Catalog Rules; 2. Accession Book Rules; 3. Shelf List Rules

ISBN/EAN: 9783337169923

Printed in Europe, USA, Canada, Australia, Japan

Cover: Foto ©Lupo / pixelio.de

More available books at **www.hansebooks.com**

Library School Rules

BY

MELVIL DEWEY M. A.

Director New York State Library; President American Library Association

Third edition

BOSTON

Library Bureau 146 Franklin St.

1899

" If the soul of the library be the librarian, its heart is the catalog."

Library School
Card Catalog Rules

with 52 fac-similes of sample cards

for author and classed catalogs

MELVIL DEWEY M. A.

Director New York State Library; Secretary American Library Association

WITH

Bibliography of catalog rules

by

MARY SALOME CUTLER

Library examiner, University of the State of New York

Third edition revized, with marginal alternative rules

Library Bureau 146 Franklin St.
1890

Note to revized edition

This edition differs from that of 1888, in containing marginal explanations, amplifications and variations from the previous rules, for which there is good authority. Variations are printed in italics, explanations and amplifications in roman type. Sample cards follow the text, not the marginal notes.

Any criticisms or suggestions for making these rules more useful will be gladly received and carefully considered before printing a third edition, if sent to

MELVIL DEWEY,

Director State Library,

Albany, N. Y.

Library School
Card catalog rules

These are the A. L. A. condenst rules made clearer on many points by re-wording and adding illustrations. They are for a double catalog of authors and markt titles on small cards (5 x 12½ cm), and of subjects on standard P size (7½ x 12½ cm). The changes are very slight in adapting the rules for a simple author catalog. For a dictionary catalog, there must be used a code of rules for subject headings.

To save detailed comparison we note that, except enlargements, these differ from the A. L. A. rules as printed some years ago, only in:

1e. We enter always under real name, omitting the exception that some books may go under pseudonyms.

1s. We follow the rule recommended as best in Cutter's rule 40, putting under the name of the place, local and municipal societies, *tho the corporate name may* not begin with that word.

2f. We give cities in their vernacular form instead of in English.

5m. We do not capitalize common nouns in German, but follow the rule of the Library of Congress.

4a. We give place and date at the end of the imprint entries insted of after edition, thus following the L. A. U. K. and Bodleian rules, the A. L. A. minority report, and the Library of Congress in putting these most important items in the most prominent place; insted of burying them back of minor items.

4c. We give edition in English rather than in language of the title.

9b. We use arabic figures for all numerals, unless roman are used on the title-page after names of rulers and popes.

The rules are in groups as follows:
  0. Check Marks.
  1. Main Entry.
  2. Heading.
  3. Title.
  4. Imprint, Contents and Notes.
  5. Capitals, Spacing and Underscoring.
  6. Arrangement of authors and titles.
  7. Subject Cards.
  8. Colored Cards.
  9. Miscellaneous.
S. C. refers to appended sample cards illustrating the rules.

o. — CHECK MARKS.

Put ORDER CHECKS on inner margin of first recto after title of v. 1.

a. "Not a duplicate." Initials of collator on order slip; or, if no order slip, as first check on book itself. If a duplicate, write "dup" and initials on book plate, if there is one; if not, in the upper left corner of front cover.

b. Source and cost of book in cents without $ sign (i. e. 145 not $1.45) after initials of duplicate collator, or, if none, as first check.

c. Pin hole in round part of last 9 in pagination shows that all order department routine is finished, and marks ownership.[1]

d. Dot under first figure of accession number shows that all entries except call numbers are made on the accession book.[2]

TITLE–PAGE CHECKS. See sample title-pages at end of S. C.

Checks under first letter of words mean:

e. ... main entry and joint authors.

f. — main entry for an anonymous book. Begins on second line of card.

g. .. added entry under editor, translator, commentator, publisher, title, etc.

h. . analytic author entry.

i. x cross reference from other forms of name, pseudonyms, initials, etc.

Other checks mean:

j. ⌐ ⌐ omit on all cards.

k. ⌐ ⌐ omit on author card.

l. ( ) series note. To follow imprint entries in ( ). S. C. 31.

m. No checks are made for biographical and subject entries.[3] Other processes are checked as follows:

n. "Cards written." Abbreviated date and cataloger's initial on first recto, after cost; and initial, with number of cards written, on bottom of main author card, the number of small cards at the left, and of large cards at the right of the guard hole. S. C. 2.[4]

o. "Cards revised." . above first letter of title-page.

p. "Classified." Class number on book-plate.

q. "Shelf listed." Book numbers on plate; and if for loan department, book card written and pocketed.

Margin notes:

[1] Oc. Each library selects its own private mark. The "last 9" given in the rule is merely to suggest the method. Some libraries use always the same page; e. g. 49 or 33 or 65. This should be a page in the first 100, that it may apply to thin books. Others choose a more complicated rule, less easily discovered; e. g. the highest page made all of 3s, i. e. 3, 33 or 333, or the highest page in 1, 2, 3 order, i. e. 12, 123 or 1234.

[2] Od. A dot on accession book against last number used, in center of class number column, shows that this number and all previous ones have been assigned.
These checks are used when accession numbers are stamped on an invoice of books before entries are made in accession book.

[3] Om. Or indicate entries for *criticism, biography and bibliography* by underlining on main cards. See Rule 8, marginal note.

[4] On. Or use *initial, without number of cards written, on back of main author card, above the guard hole.* Always write entries on the back of the card so that they can be read from the front of the drawer by simply tipping the card forward.

4

ꝏ *r.*   "For inspection."   Thread of proper color for each day.[1] M. Blue.   Tu. Green.   W. White.   Th. Red.   F. Fawn.   St. Salmon.

ꝏ *s.*   "Gilded."   Numbers on backs.

Checks a–d are made by order clerk; e–l by head cataloger as guide to carders.[2] All the checks except e–l, are made by the person doing the work, as a record, and must never be put on till it is finished.

1 Or.   *Or colored card in pocket or between leaves with name of day and explanation printed on it.*

2   *Or as record of work done; omit j–l if title pages are not markt for carders.*

I. — MAIN ENTRY.

Enter books under :

a.   Authors' SURNAMES if known. S. C. 1 and 2. ꝏ *except ir no. s.c. ꝏ*

b.   EDITORS OF COLLECTIONS, S. C. 4 (also cataloging each separate item, if expedient). S. C. 24 and 25.

c.   COUNTRIES, CITIES, SOCIETIES, LIBRARIES, or other bodies responsible for their publication. S. C. 5.[3]

d.   Authors' INITIALS, when these only are known, putting last initial first. S. C. 18. Make also added entry under title. When author's name is found fill it in on all cards and write a new card refering from initials to full name. S. C. 19.

3 1c.   *After names of all except very prominent cities add the name of country or state in ( ); e. g. Boston (Eng.), Utica (N. Y.); also, if necessary to distinguish, add (state),(city) or (county); e. g. N. Y. (state).*

e.   PSEUDONYMS when real names cannot be found. Add *pseud.* 1 cm after, if sure the name is not real. S. C. 20. When the real name is found write it above the pseudonym on the cards (or re-write cards). S. C. 21. Refer from pseudonym to real name. If there is only one book under that pseudonym, its title and call number may be given on the card, S. C. 22, but, if there are several works, the real name only is given, S. C. 23,[4] as the card must contain all the titles in the library or none.[5]

4 1e.   *Or in case of prominent pseudonyms, like George Eliot, add brief titles and call numbers of all the works, giving one line to each entry.*

5 1e.   *Make added title entries for all pseudonymous books.*

f.   FIRST WORD (not an article or serial number and omitting mottos or designations of series) of titles of anonymous books whose authors are still unknown, S. C. 15, and of periodicals. S. C. 28 and 29. (A book is anonymous when the author's name is not on the title-page, — see Cutter's Rules p. 10,[5*]—or in government publications, on pages following the title-page.) When author's name is found, pencil it on title-page and enter on author line of cards, putting [ ] around the name, and *anon.* 1 cm after, S. C. 16, making also a new card with added entry under title followed by author's name in [ ]. S. C. 17.[6]

5* 1f.   *no of ed. s*

6 1f.   *Or for anonymous biographies whose author is still unknown, make given biography card the main card for author catalog. (See 2b, N. t', 40), and make added title entry if merited.*

Enter :

g.   COMMENTARIES with text, and TRANSLATIONS, like the original, with added entry under commentator or translator; S. C. 6, 7, and 8, but commentaries without the text, under commentator only, tho classed with the original ; e. g. Coleridge's "Notes and lectures

5

upon Shakspere" is entered under Coleridge, tho classed with Shakspere.

h. BIBLE, or any part of it, including the Apocrypha, under editor, translator, etc. ; e. g. Alford, H : ed. Greek testament with revised text ; entered under Alford in form of S. C. 4, or, if no editor, under the first word of title, writing no author card ; e. g. "Das neue testament" to be entered like an anonymous book. S. C. 15. Most libraries enter all bibles under the word Bible. This would duplicate in the author catalog the group under Bible in the subject catalog, to which reference is made from the word Bible in the author catalog. If *only* an author catalog is made, it should include all bibles under that word.

i. TALMUD, KORAN, VEDAS, and other sacred books under those words on author line, making added entries under editor, translator, &c. S. C. 34.[1]

I II. In like manner enter other sacred or well-known anonymous books under their best known title; e. g. Beowulf, Nibelungenlied, Roland.

j. ACADEMICAL THESIS under respondent or defender, unless præses is clearly the author.

k. Books having MORE THAN ONE AUTHOR under the first named in title, with added entries under each of the others. S. C. 13 and 14. If only two or three authors, include their names in the main entry ; e. g. "Roe, R :, Doe, J : and Smith, D :" ; if more than three, enter under the first "and others " ; e. g. "Roe, R : and others," and make an added entry under each of the others ; e. g. "Doe, J :, Roe, R : and others," "Smith, D., Roe, R : and others." Write each author's full name only once and that on his own author card. S. C. 13 and 14. Joint editors, translators, etc., have separate added entries. S. C. 9. Parties in a debate are treated like joint authors. official & anon reports

TRIALS of crown and criminal cases under defendant; e. g. Burr, Aaron, Trial for treason ; civil cases under the parties in the suit, treated like joint authors ; e. g. Vanderbilt, Cornelius, vs. Livingston, J. R.; marine cases under the ship ; e. g. Blaireau (ship) ; PLEAS, DECISIONS, etc., under the author, with added entries as below. A plea printed separately goes under the lawyer making the plea, with added entries under the party or parties in the suit ; e. g. Smith's argument in case of prize steamer Peterhoff, is entered under Smith, with added entry under Peterhoff. Decisions published separately go under court, with added entries for parties in the suit.

m. CATALOGS of private collections under owner ; catalogs of public collections by rule 1 c. Make added entry under compiler, e. g. Daniel, G: Catalogue of [his] library ; Boston Athenæum, Cata-

6

logue of the library [by C: A. Cutter]; with added entry under Cutter.

✓ *n.*   ALUMNI proceedings, etc., and local COLLEGE SOCIETIES under the college; e. g. Yale college — Skull and bones society. Enter their publications under heading above with reference from Skull and bones society. Chapters of FRATERNITIES under name of fraternity, with added entry under the college; e. g. Psi Upsilon fraternity.

✓ *o.*   NOBLEMEN under their[1] titles with reference from family name.. If the family name[2] is decidedly better known, enter under that with reference from title; e. g. Romney, H: Sidney, 1st earl of; S. C. 36, but (the only exception for Englishmen), Bacon, Francis, viscount St. Albans.

✓ *p.*   ECCLESIASTICAL DIGNITARIES, unless popes or sovereigns, under their surnames, e. g. Newman, J: H: card. Butler, Jos. bp.

✓ *q.*   SOVEREIGNS, e. g. Napoleon 1, (except Greek and Roman, e. g. Justinianus 1, Flavius Anicius); ruling princes; Oriental writers; popes, e. g. Leo 13; friars, e. g. Hyacinthe, Père (C: Loyson); persons canonized, e. g. Ambrosius, St., and all others known only by their first names, under the first name.

✓ *r.*   MARRIED WOMEN, and other persons who have changed their names, under the last well-known form, with reference from other forms. Enter Helen Hunt under Jackson, Mrs Helen Maria (Fiske) Hunt, with references from Hunt, and H. H.[3]

✓ *s.*   A SOCIETY, UNDER FIRST WORD (not an article) of its corporate name, with reference from any other name by which it is known, specially the place if it has head-quarters and is often called by that name, e. g. Statistical society of London, with reference from London, statistical society: UNDER PLACE, local societies, e. g. N. Y. Shakespeare society; academies of the European continent and South America, e. g. Berlin akademie der wissenschaften; municipal institutions, viz. libraries, e. g. Boston — Public library; galleries, e. g. N. Y. Metropolitan museum of art; public schools, e. g. Cincinnati — Education, Board of; and municipal corporations, e. g. Minneapolis — Health, Board of: UNDER THE STATE, state historical societies and state colleges, e. g. Wisconsin historical society, California university.

✓ *t.*   CYCLOPÆDIAS, DIRECTORIES and ALMANACS, under title, like a periodical, S. C. 28 and 29, making added entry for editor, publisher, partial title, or any form under which they may be well known, e. g. enter Appleton's cyclopædia under American

7

encyclopædia with added editor entry for Ripley and Dana,
and a similar one for Appleton, pub. But if it is the work of
an individual, e. g. Larousse, enter under author's name with
added entry under title.

✔ *u.* A PERIODICAL which is the organ of a society or club, under its
name, like S. C. 28 and 29, with added entry under name of
society, unless it be the regular proceedings or transactions, in
which case enter it under the society, with added entry under
title, like S. C. 11 ; e. g. Library journal as periodical, with added
entry under A. L. A.; but Quarterly journal of the Geological
soc. of London, under the society's name. In case of change of
name of periodicals, enter the whole under the earlier name,
making note of the change on the main cards. Make added
entry under the new name, giving imprint of the part covered by
it, referring on the same card to the old name.

v. SERIES under editor, if known, S. C. 30, with reference from
title ; if unknown, under title, e. g. Bampton lectures.[1] For some
series a series card need not be written, e. g. Clarendon press
series. Give two lines to each item of contents on series card,
beginning with series number between red lines, thus giving
space for both class and book numbers. S. C. 30.

w. A BIOGRAPHY under its author, putting the full name of its subject
on the upper line of the subject card, followed by the dates of
his birth and death. S. C. 26.[3] See also 7 f.[4] If author's name
is not known, leave second line blank, and fill in if found later.
An autobiography, e. g. Grant's Memoirs, is written like a
simple subject and author card, like S. C. 1 and 2, except that
the full name and dates are added on the subject card ; e. g.
Grant, Ulysses Simpson, 18th pres. of U. S. 1822–1885.
Diaries and letters are considered autobiographical. When it is
found that authorities differ, affix ? to the date used ; but unless
in special cases do not attempt to settle the discrepancy. If the
choice is between only two dates, both may be given, e. g. b.1844
or 6. Where exact dates are unknown the time when the
author flourished may be given, e. g. "fl. 13th cent." or "fl.
1430 ? " using the ? instead of *circa* or *about* for approximate dates.[6]

x. Make added entries, called ANALYTICALS, for distinct parts of books;
enter under author of part analyzed, and give brief title and
date of main work. See 4r. S. C. 24 and 25.

y. Analyze a LIFE INCLUDED in another work by an author card for
the author of the life and a subject card for its subject. S. C. 27.[5]
The writer of an "included," who also edits the whole work, has
no author analytical, as the editor card is enough.

*z. Make ADDED ENTRIES under titles of all novels and plays, and all other striking titles. S. C. 10 and 11 ; specially noticeable words in titles, S. C. 12 ; translators, commentators, noted illustrators, editors of books and periodicals, S. C. 6, 7 and 8 ; and make REFERENCES from pseudonyms, initials, and from an ecclesiastical title when that and not the family name appears on the title-page. (e. g. [Andrews,] Lancelot, bp. of Winchester. Since Andrews does not appear on the title-page, make reference from Winchester.) If a book is well known by any title (e. g. running, half, or binder's title) differing from the title-page, make added entry under this other title. S. C. 12. If an author is commonly known by a name less full than that used for the main entry, refer from the shorter to the full entry, e. g. make Müller, F; Max, the main entry, and write a reference card thus :— Müller, Max, see Müller, F; Max.[1] Make added entry or reference in every other case when needed for the ready finding of the book.

2. — HEADING.

*a. Give author's NAME IN FULL, and in the VERNACULAR, with reference from any other form by which it is commonly known; but give the *Latin form* when better known, with reference from the vernacular; e. g. Estienne, H:, with reference from Stephanus, but Grotius, Hugo, with reference from Groot, Hugo van. S. C. 35.[2]

    Enter sovereigns under the English form ; e. g. William (not Wilhelm) 1, king of Prussia.[3]

    In heading and notes if there is equal authority for more than one SPELLING, prefer the shorter ; e. g. abate, not abbate, Shakspere, not Shakespeare.[4]

    For full names and dates of Greek and Latin authors follow Smith's Dictionary of Greek and Roman biography.

*b. On subject cards write out author's FORE-NAME if only one, e. g. Phillips, Wendell ; if more than one give simply initials,[5] e. g. Beecher, H: W., except in case of autobiography. See 1 w. The colon abbreviations are used for both full name and initial, H: meaning Henry and being just as brief as H. in giving initials. On author cards give author's full name either written out or in the "Library abbreviations," e. g. Emerson, Ralph Waldo, Beecher, H: Ward, Allen, Jos. H:

*c. Enter English and French surnames beginning with PREFIX (except the French de and d') under the prefix ; in other languages under the word following, e. g. La Fontaine, Voltaire, F. M. A. de, Göthe, J; W. von, Lennep, Jacob van, but enter foreign names anglicized under prefix, e. g. De Peyster, Da Ponte, Van Nostrand.   **9** *d.*   Enter English COMPOUND NAMES under the last part, foreign under the first, with reference from form not chosen, e. g. enter S. Baring-Gould, under Gould, with reference from Baring-Gould, but E. P. Dubois-Guchan under Dubois-Guchan, with reference from Guchan.

e.   Add designations (such as TITLES OF HONOR, DATES OF BIRTH AND DEATH, RESIDENCE, etc.) to *distinguish* writers of the same name. Omit the article with such added titles, e. g. abate not il abate, abbé not l'abbé. Always add titles that are part of the person's usual designation, as Rev., Gen., etc. ; or a sobriquet, e. g. Jackson, T: Jonathan, called Stonewall, Stewart, J: called Walking. Do not bracket such added titles. Distinguishing dates follow title affixes, and precede the affixes, ed., tr., etc., e. g. Smith, W: LL.D. 1814– tr. Anon. follows distinguishing dates and the words ed., tr., etc.

> Dates of birth and death are so compact and useful a distinction that they should be put on the cards *in all cases* if the cataloger has them before him, and should be *looked up* whenever there is question of identity between similar names.

f.   Give names of CITIES AND TOWNS in the vernacular, but larger political divisions in English, e. g. Wien, not Vienna, (with reference from English form), but Austria, not Österreich.

g.   When author's name is not in title of first volume, but is in that of any other, make both author and title entries, S. C. 10 and 11, but do not enclose author's name in [ ] nor add *anon.*

h.   To avoid confusion with author's initials, etc., letter to resemble *print* (2-3 size) titles of honor and similar distinguishing words when they precede the fore-name, also letter-titles affixed, as D. D S. C. 23.

i.   On ADDED ENTRY cards, write author's name on the second line, with full FORE-NAME if only one, with initials if more than one, like the form for heading on subject cards. Rule 2b. S. C. 8, 9, 11, 12, 22, 33. On REFERENCE CARDS, write author's name on the second line, in full. S. C. 19, 23, 35, 36.

j.   For GOVERNMENT DEPARTMENTS &c. use the official name, inverting, if necessary to alphabetize by the distinctive word ; e. g. France — Travaux publics, Ministère des. S. C. 5.[1]

k.   In the often ABBREVIATED ITALIAN NAMES use the full form ; e. g. Giovanni Battista, not Giovambattista, or Giambattista ; Giovanni Giorgio, not Gian Giorgio ; Antonio Francesco, not Antonfrancesco ; Pietro Francesco, not Pier Francesco.

1 2].   *or enter subordinate bureaus directly under the bureau, and under the department give a list of all bureaus belonging to it under which entries have been made.*

2 2].   Bracket all names and parts of names not found on the title-page, except in red ink subject headings. S. C. 26-27, 38, 40-41, 43-44, 46.

10

3. — TITLE.

a. Make title an EXACT COPY of title-page, neither corrected, translated, nor in any way altered; but *omit* mottos, honorary titles,
repetitions, and fore-names except initials of editors, translators,
etc., and any other matter not essential. Indicate all omissions
on main cards (except initial article in English, and author's name
and titles in usual position), by *three dots* ... which take the
place of all punctuation. S. C. 5, 6, 13, etc.[1] Titles of books
specially valuable for antiquity or rarity give in full, with all
practicable precision. Copy exactly the phraseology and spelling
of the title. For spelling of heading see 2 a. Punctuation
must be supplied if omitted on the title-page and must sometimes be changed on account of omissions or to reduce the
title to a single sentence.

b. Indicate MISPRINTS, or odd spellings, by three dots underneath,
rather than by the common [*sic*], e. g. Kanzas not Kanzas [*sic*].

c. Follow old interchange of U AND V, I AND J, only in books before
1600 A. D. S. C. 6 and 7.

d. PUNCTUATE by Cutter's Rules, No. 163, and Bigelow's Punctuation.

e. Supply in [ ] in Eng. any ADDITIONS needed to make the title clear,
e. g. Examination of the president's [T : Jefferson] message.[2]

f. Do not translate into the vernacular proper names occurring in
FOREIGN FORMS in the title, e. g. "ed. a J. G. Amesio," not "ed.
a J : W : Ames," but give the vernacular form as the heading of
the reference card. S. C. 8.

g. TRANSLITERATE by A. L. A. rules, titles in foreign characters, except Greek. If the title does not show it, state language of the
book in [ ]. S. C. 32.[3]

h. In case of two title-pages in different languages, use the language
in which the book is written, if that has the Roman alphabet;
otherwise use the title-page having the Roman alphabet, e. g.
when a book has both GREEK AND LATIN TITLES, give the Latin.
If the book itself is in two languages and has two titles in
the Roman alphabet, use the first. ACCENT words in Greek,
French, etc., regularly, though the title-page be all in capitals
without accents. S. C. 10 and 11.

i. After title specify APPENDIX of single volumes, e. g. [apx. p. 320-
346], but only in exact work.

j. Mention frequency of publication of PERIODICALS, e. g. Contemporary review; [monthly]. Do not repeat if mentioned in the
title. S. C. 28 and 29.

k. Begin PERIODICALS on author cards at left-hand red line for first
two lines, using full lines for the rest; on subject cards, begin at

11

right-hand red line for first line and left-hand red line for the rest. In both cases begin at top line. S. C. 28 and 29.[1] Anonymous books are to be entered on second line, filling in author's name when found, on first line. S. C. 15 and 16.

l. Omit INITIAL ARTICLE in English except in a few cases where its omission would destroy the sense; e. g. The judgment, a vision. In other languages express the article, but alphabetize by following word. Inclose initial articles in parentheses.

m. REPEAT the AUTHOR'S NAME in the title in those few cases where it is an integral part of the title; e. g. Monteith's geography.

4. — IMPRINT.

a. Order. series of periodicals, etc. (in English). For other " series " see 4s.
edition (in English).
pages (or volumes if more than one).
illustrations.
group of portraits. ⎫
portrait of a group. ⎬ Use only in exact work.
portraits.
plates.
photographs. Use only in exact work.
maps.
fac-similes.
tables.
size (by letter; maps, broadsides, etc., hight x width, in centimeters, e. g. 91 x 71 ᶜᵐ).
no title-page (if there never was any).
place.
publisher's last name, in books before 1600 A. D. S. C. 6 and 7.
date.
copyright date, if differing more than a year from date of publication. S. C. 3.
For full illustration of imprint see S. C. 3.

b. Give *initial capital* to first word of imprint unless the imprint begins with a number. S. C. 26 and 34.

c. Give EDITION in English in all cases,[2] and omit all adjectives except new, the number of the edition, and enlarged;[3] e. g. not 2te durchgesehene, vermehrte und verbesserte auflage but Ed. 2 enl.

Give extremes of *various editions*, e. g. Ed. 1—4.

d. Give numbers of PAGES, connecting the last number of each paging with the sign + and adding unpaged matter in [ ]; but where

12

there are four or five pagings, add the shorter so as not to have

[handwritten marginal notes]

number of pages; e. g. 16 p. + 3202 col. [801 p.].

If two or more volumes are PAGED CONTINUOUSLY, give number of volumes and pages in imprint; e. g. 2 v. 21 + 360 p., and add in a note " Paged continuously."

e. When VOLUMES of a set are missing, give number of volumes in complete set ; and specify missing volumes in pencil note, which can be erased as soon as they are secured. S. C. 28 and 29.

f. In CURRENT PERIODICALS and continuations, give exact statement of volumes in library if the library contains less than half what has been published; e. g. Atlantic monthly, 1858–75, 1880–81, 1884–date v. 21–36, **v.** 45–48, v. 53–date; if it contains more than half, give in imprint statement of entire work, and missing volumes in note. S. C. 28 and 29. Of a set published at intervals, like Stephen's "Dictionary of national biography," give in imprint in pencil exact statement of what is in the library, correcting pencil entry on receipt of each new volume, and entering in ink when complete.

g. If a book is INCOMPLETE and no more will be published, give in the imprint what has appeared, adding in a note "no more published," e. g. American archives.

h. Disregard ENGRAVED AND ILLUSTRATED TITLE–PAGES except in exact work. If no title-page was published, write in imprint n. t-p. [no title-page]. S. C. 3. If the title-page is otherwise missing, add in note t-p. w. [title-page wanting]. S. C. 13.

i. Except in exact work, give number of PLATES, PORTRAITS, ILLUS-

13

right-h

rest.

mous

1 3k. Or if author cards are written on P size (7.5x 12.5 cm) indent like subject cards.

name

Omit ini

omiss

In o

follo

m. REPEA

it is

a. Order.

illustrations.
group of portraits. }
portrait of a group. } Use only in exact work.
portraits.
plates.
photographs. Use only in exact work.
maps.
fac-similes.
tables.
size (by letter; maps, broadsides, etc., hight x width, in
centimeters, e. g. 91 x 71 cm).
no title-page (if there never was any).
place. "Various places" sometimes used
publisher's last name, in books before 1600 A. D. S. C.
6 and 7.
date.
copyright date, if differing more than a year from date of
publication. S. C. 3.
For full illustration of imprint see S. C. 3.

b. Give *initial capital* to first word of imprint unless the imprint begins
with a number. S. C. 26 and 34.

c. Give EDITION in English in all cases, and omit all adjectives except
new, the number of the edition, and enlarged; e. g. not 2te
durchgesehene, vermehrte und verbesserte auflage but Ed. 2 enl.

2 so. or give in vernacular if it is copied as a part of the title.
1 so. translating equivalent phrases into these;

Give extremes of *various editions*, e. g. Ed. 1—4.

d. Give numbers of PAGES, connecting the last number of each paging
with the sign + and adding unpaged matter in [ ]; but where

12

there are four or five pagings, add the shorter so as not to have more than three groups of figures, usually the main paging preceded and followed by the totals of the shorter pagings. If there are more than five pagings do not count, but write v. p. [various paging], except in exact work. If the book is *unpaged*, write unp. and for valuable or very thin books count and add pages in [ ]. Give paging of *last recto* when its verso is unpaged, but last recto unpaged, if printed is + [1]. S. C. 3.

Disregard ADVERTISING pages except when paged in consecutively. Do not specify folded leaves paged with the rest, except in exact work.

Accept the paging of the book unless there is an obvious misprint, e. g. the first page may be numbered 5, but do not deduct the four omitted pages. In books which are only portions of other books give inclusive pages; e. g. p. 633 — 742.

Write f. (or leaves) instead of p. where the book is *foliod*, i. e. numbered by leaves instead of pages; e. g. 75+260 f.

If columns not pages are numbered, give columns and in [ ] number of pages; e. g. 16 p. + 3202 col. [801 p.].

If two or more volumes are PAGED CONTINUOUSLY, give number of volumes and pages in imprint; e. g. 2 v. 21 + 360 p., and add in a note "Paged continuously."

e. When VOLUMES of a set are missing, give number of volumes in complete set; and specify missing volumes in pencil note, which can be erased as soon as they are secured. S. C. 28 and 29.

f. In CURRENT PERIODICALS and continuations, give exact statement of volumes in library if the library contains less than half what has been published; e. g. Atlantic monthly, 1858–75, 1880–81, 1884–date v. 21–36, v. 45–48, v. 53–date; if it contains more than half, give in imprint statement of entire work, and missing volumes in note. S. C. 28 and 29. Of a set published at intervals, like Stephen's "Dictionary of national biography," give in imprint in pencil exact statement of what is in the library, correcting pencil entry on receipt of each new volume, and entering in ink when complete.

g. If a book is INCOMPLETE and no more will be published, give in the imprint what has appeared, adding in a note "no more published," e. g. American archives.

h. Disregard ENGRAVED AND ILLUSTRATED TITLE-PAGES except in exact work. If no title-page was published, write in imprint n. t-p. [no title-page]. S. C. 3. If the title-page is otherwise missing, add in note t-p. w. [title-page wanting]. S. C. 13.

i. Except in exact work, give number of PLATES, PORTRAITS, ILLUS-

13

TRATIONS, etc., only when on the title-page or quickly ascertainable.

Portraits, plates, maps, etc., included in the regular paging are simply illustrations.[1]

j. If there is an ATLAS or volumes of PLATES, write, e. g. v. 1 — 12 O, and Atlas F, or if they are of the same size, v. 1 — 12 and Atlas O.

k. If volumes of a set have DIFFERENT SIZES specify volumes of each size; e. g. 12 v. (v. 1 — 4 O, v. 5 — 12 Q).

If a bound volume of pamphlets has parts of different sizes, give actual size of each part on its own cards ; in accession book give size of outside of bound book.

l. Give first PLACE OF PUBLICATION, unless another is *known* to be the true place, or the book actually was printed in more than one place, when both are to be given.

m. Give EXTREME DATES where the volumes of a set differ, and also date of v. 1. if later than first date ; e. g. 1834 — 49 [v. 1, '38]

If the copyright dates also differ, write, e. g. 1834 — 49 [v. 1 '38] [°1824 — 31]. Give all these peculiarities of date only on the main author and subject cards. *& col. cds cnh.*

n. Give PLACE AND PUBLISHER'S NAME in language of title. S. C. 6 and others ; corrections and additions in English, enclosed in [ ] ; e. g. Camb. [Eng.], or Camb. [Mass.] Give publisher's name for Bibles. *& books pub. before 1600.*

o. Use *only* the ABBREVIATIONS in the printed "Library abbreviations compiled by Melvil Dewey."

p. Write FULL IMPRINT only on the main subject and main author card. For joint authors, and other added entries, the date is sufficient. S. C. 13 and 14.

CONTENTS AND NOTES.

q. Give NOTES (in English) and CONTENTS of volumes in smaller letters (generally only on subject card), when necessary properly to describe the work. Notes about the author and on imperfections in the copy go on both cards, other notes only on subject card.[2] Begin notes, except note abbreviations, with capitals, starting on second line after imprint and indenting like title.

r. In ANALYTICAL REFERENCES when the article is independently paged, give full imprint of the analyzed part, and in ( ) write "in" and title of main work. S. C. 24. If paging is not distinct, write in ( ) "see," and title, and date of main work, with pages where analyzed matter is found. S. C. 25.

s. SERIES NOTE follows date on main cards in ( ).[3] Always specify

---

Marginal notes:

[1] 4 i. In serials and transactions do not distinguish portraits, plates, etc., but use simply "il." and this only when the serial is usually illustrated.

[2] 4 q. Or other notes generally on subject card only.

[3] 4 s. For fullness of author's name follow rule for subject headings. See 2b.

volume of series if known. S. C. 31. For "series" of periodicals, etc., see Imprint 4a.

✓t. In contents and series begin each separate entry on a new line and keep all entries to the right of the right hand red line, putting volume numbers, if they occur, between the red lines.[1]

✓u. Note MS ANNOTATIONS if they seem to add to the value of the book.

[1] For fullness of author's name follow rule for subject headings. See 2b.

5. — CAPITALS.

These are substantially the rules adopted by the Harvard college library:

· In titles, notes, and whatever goes on the body of a card, capitalize as follows:

✓a. The FIRST WORD of every sentence, of every title quoted, S. C. 15 and of every alternative title introduced by *or*. S. C. 10. In quoting titles like the Nation, the Times, etc., capitalize the word following the article and not the article, and do this even in defiance of quotation marks ; e. g. extracted from the Times, extracted from "the Nation." This rule allows capitals to the Bible, the Scriptures, the Book of Mormon, etc.

✓b. Names of PERSONS.

✓c. EPITHETS standing as substitutes for PERSONAL names : e. g. the Pretender. The epithets, His Majesty, Sa Majesté, His Excellency, etc., when not followed by the personal name or by the titles king, president, etc., are substitutes for a personal name, and should be capitalized. But when followed by the personal name, or by the title, such epithets should usually be omitted ; e. g. "the presence of His Majesty at that time," "the coronation of . . . George III.," "the favor of . . . the king." When these epithets occur with superfluous adjectives the latter should be omitted ; e. g. not "His Most Glorious Majesty," but "His . . . Majesty." The rule allows capitals to Trinity, the Deity, the Creator, etc., but do not capitalize holy, sacred, divine, etc., except in Holy Ghost, Holy Spirit.

✓d. Mr, Mrs, Miss, Dr, Sir, Lord, Lady, Monsieur, Madame, Mademoiselle, Signor, Don, Herr, Frau, used as PREFIXES to names of persons.

✓e. The Great, the Lion-Hearted, le Grand, der Grosse, etc., used as AFFIXES to names of persons.

✓f. Names of PLACES. These often consist of an individual name joined to a generic name. In such cases capitalize only the former ; e. g. state of Connecticut, Berkshire county, city of Boston, Susquehanna river, Catskill mountains, Arctic ocean, south Pacific, east Tennessee, tropic of Cancer, Arctic regions, equator. But there are some cases in which the generic name has come to be

15

so closely united with the individual name that both should be capitalized; e. g. Niagara Falls, White Mountains, Mont Blanc, Lake Erie, Zuyder Zee, North Carolina, Lundy's Lane, Van Diemen's Land, North Pole, Bull Run, Fall River, Mound City, the steamer "City of Boston," etc. It is not generally difficult to distinguish between these two cases. Ability to use the individual name by itself will usually afford a safe criterion; e. g. we can say "the Catskills," but not " the Whites."

√ *g.* EPITHETS standing as substitutes for names of PLACES; e. g. the South, the Orient, United Kingdom, etc.

√ *h.* Arbitrary, undescriptive, fanciful, outlandish, or otherwise purely INDIVIDUAL EPITHETS occurring in the name of a society, corporation, or building; e. g. Vulture insurance company, Pi Eta society, Globe bank, Star and Garter inn, Adelphi, Star chamber, Excelsior mine, Court of Oyer and Terminer, Chrestomathic day-school, Old Bailey. Capitalize names of societies or collective bodies, and write Royal society, Board of trade, House of representatives, First congregational church, Harvard college, American academy of arts and sciences, State department, Oxford university, Parliament, College of physicians and surgeons, etc.

√ *i.* The pronoun I; the interjection O; A. D., B. C., in dates; D. D., M. D., etc.; but not ms., mss.

√ *j.* In English and Latin, but not in other languages, capitalize also ADJECTIVES derived from names of PERSONS and PLACES; e. g. English, Platonic, etc.[1] This rule allows the capitalization of many names of parties and sects which may be regarded as adjectival nouns derived from proper names, as Lutheran, Arminian, Jesuit, Christian, Buddhist, etc. Otherwise do not capitalize such words; e. g. catholic, episcopal, puritan, whig, democrat, quaker, unitarian, etc.

√ *k.* Names of the MONTHS, DAYS of the week, and HOLIDAYS, but only the individual part of the name; e. g. Shrove Tuesday, Candlemas, 4th of July, Fast day. Capitalize also Advent, Lent, Lord's Supper.

√ *l.* Pope, Saint, Bp., King, Earl, Capt., Rev., Hon., Prof., Judge, Gov., etc., used as PREFIXES to names of persons; e. g. King George III., Earl Russell, Bp. Colenso, Secretary Fish. Otherwise do not capitalize such words; e. g. the king of England, the earl of Derby, the bishop of Lincoln, the secretary of war.

√ *m.* Except in the cases specified above, use SMALL LETTERS exclusively, tho local usage require capitals, e. g. nouns in German. But do not use capitals if local usage is small letters, e. g. names of months in French.

[1] *j. Or capitalize adjectives and nouns derived from names of persons and places, unless reputable usage justifies lower case.*

16

SPACING AND UNDERSCORING.

n. Leave SPACE of one centimeter in *heading*, between author's name and words or dates affixed, but write dates on biography cards at the end of the line. S. C. 26 and 27.[1] The first line (or more if needed, S. C. 13 and 14) is reserved for the author's name, and is left blank if the author is not known. S. C. 15.

The title begins on the next blank line tho there be vacant space on the author line.

But periodicals are entered on the author line. S. C. 28 and 29.

o. Leave also one centimeter between *title* and *edition*, between edition and the rest of the *imprint*, between *size* and *place*, and between date and series note, or other matter following date on the same line.[2]

p. On main cards SINGLY UNDERSCORE secondary entries and references ; on secondary cards DOUBLY UNDERSCORE main entry, but do not underscore secondary entries or references.[3] S. C. 13 and 14. For *joint authors* of series singly underscore the first in series note on main cards, and the others on series card.

6. — ARRANGEMENT.

a. SURNAMES when used alone precede the same names with fore-names ; initials of fore-names precede fully written fore-names beginning with the same initials, (e. g. Brown ; Brown, J. L.; Brown, Ja.).

b. PREFIXES M' and Mc, S., St., Ste., Messrs., Mr and Mrs, arrange as if written in full, Mac, Sanctus, Saint, Sainte, Messieurs Mister, and Mistress.

c. WORKS of an author arrange thus : —

1. Collected works.
2. Partial collections.
3. Single works, alphabetically by first word of title.

d. Alphabet in order of English alphabet.

e. German ae, oe, ue *always* write ä, ö, ü, and arrange as a, o, u ; e. g. Göthe, not Goethe. But in case of several surnames which are alike except the umlaut, arrange the umlauted after the plain letter ; e. g. all the Müllers after all the Mullers, W. Muller coming before A. Müller.

f. Names of PERSONS precede similar names of PLACES, and places precede TITLES, e. g.

Washington, G :

Washington, (D. C.).

Washington Adams in England. · (Title of book.)

See also Cutter's Rules, No. 169–196, e.r\. nf 13h3 or 13`7\, no. 213 – 258. 17

---

[1] 5n. This space may be shortened to avoid over-running the line. Or follow the same rule for dates on biography cards.

[2] 5o. If space would come at beginning of line, leave no space. On all secondary cards and in series, notes, etc. leave 1-2 centimeter in all similar places; e. g. on second line of editor, translator, biography cards, etc. and between title and date, title and volume, etc.

[3] 5p. unless they do not appear on main cards.

7.—SUBJECT CARDS.

a. In general write subject cards like main author cards with author's name on upper line, followed by title and imprint. Name of subject entry is not written (except where divisions more minute than in the printed classification are introduced alphabetically under final class no. see 7 m), but subject is shown by class number in upper left corner of card. Note the following points of difference : —

b. SIZE. Use P size [7.5 x 12.5 cm] for subject cards, I size [5 x 12.5 cm] for author and title cards. {This distinction in size is made at Columbia, but is not essential. Many prefer to use P size for all cards. See Library notes 1: 34⁶.]¹

c. FULLNESS OF ENTRY. On subject cards abbreviate author's forename and give a fuller title. Rule 2b. S. C. 1, 6, etc.

d. MARGIN. On subject cards leave all space at left of first vertical line free for stamping cross reference subject numbers.

e. PERIODICALS. Indent subject card of a periodical by Rule 3k. S. C. 28 and 29.

f. BIOGRAPHY. Add name of subject of life with date of birth and death on upper line.² See Rules 1w, 1y, and S. C. 26, 27, and 38.

g. STAMP call number of every book (thus indicating both its subject and location on shelves) on *all* its cards ; class number first, and directly under this its book number, in blue ink in upper left corner of card, on first and second lines of main cards, and on second and third of cross reference subject cards. Stamp cross reference number in red ink on upper line of reference card, and on fourth, and (if more than one reference) on succeeding lines of main subject card. S. C. 27, 42, 43, 45, 47, 48, 49, 50, and 51. Put no cross reference numbers on small cards. Add vol. number in blue ink to all cards referring to only part of a set. S. C. 24, 25, and 27.

There are three forms of cross reference subject cards, each stamped according to Rule 7g.

h. GENERAL ; where the book as a whole treats of more than one subject. Make reference card for the less important subject, giving simply author, brief title, and date. S. C. 48 and 49.

i. ANALYTICAL. This may be biographical, like S. C. 27, or analysis of any other subject of importance. S. C. 50 and 51.

j. INDEPENDENT ; pamphlets or books on different subjects bound in one volume. Card each independently, i. e. subject and

---

Margin notes:

1 7b. *If P size is used for author cards, indent them like subject cards, as the larger card makes it less important to use the left margin, and there are practical advantages in having author and subject cards alike.*

2 7f. *In red ink.*

author cards, with full imprint for each part. Locate book by most important subject, and treat cards of this part as main cards, stamping subject cards for other parts as cross reference cards. Do not stamp these cross reference numbers on main card.

k. ARRANGE subject cards numerically by class numbers, and under each number by authors, disregarding book numbers ; except for colleges [378.4–9] and authors to whom a special class number is assigned in literature and philosophy, which arrange under each class number, by book numbers instead of by authors.

l. GENEALOGY. In 929.2 alphabet by subject instead of author, writing subject name, e. g. Grant family, on the top line, as in S. C. 26.

m. USE "SPECIAL TOPIC card," where an alphabetical arrangement of minute subjects is desirable under the class number ; e. g. 974.41 means history of Berkshire county, Mass. ; for history of several towns in that county, write name of each town in red ink above author's name on its subject card, thus making it a "special topic card ;" arrange these alphabetically by red ink — special topics at end of author arrangement under class number, and put a "special topic guide," S. C. 47, containing a list of these towns, at beginning of 974.41.

8. — COLORED CARDS.

To make certain classes of entries more prominent, use cards of distinctive colors for both main and analytical cards.[1]

[1] 8. Indicate these on main cards by underlining the names under whic[h] these entries are made or by a note, if necessary.

COLORED CARDS IN AUTHOR CATALOG.

This makes a *name* catalog in place of a strict *author* catalog.

a. BLUE cards for author BIBLIOGRAPHY. S. C. 46. Use these in name catalog instead of subject cards under 012 in subject catalog. Put in 012 a card with this note : — " *For bibliography of any special author see blue cards standing first under his name in author catalog.*"[2]

[2] 8a. Or let these duplicat[e] in name catalog th[e] while subject cards un[der] der 012 of subject ca[t]alog.

b. GREEN cards for BIOGRAPHY. S. C. 40 and 41. These duplicate subject biography cards (S. C. 26, 27, 38), but are arranged in the single alphabet of the name catalog, where they may be found without knowing anything but name. The subject cards show biography of any class, though readers know no names.

Jun. crit. indicated by E or D
In long controversy. Make no more than 3 yellow cards.
A BC typical form. Treat C as direct criticism of B and ⌐
another yellow card for a for author cat. without imp.

B A² make a² a direct crit. of B and no third card in this ⌐
B C D Cat. D as c.   c. YELLOW cards for CRITICISM. 1. General criticism of an author.
rit. of C + make in   S. C. 44. 2. Criticism of an individual work. Make author card
no yellow cd for a.   similar to S. C. 43.
drops crit.

    d. ARRANGE colored and white cards in name catalog: —

backup— A B C   1. Bibliography (blue).
822.33 D in this   2. Biography (green).
we make C a direct   3. Complete works (white).
rit. of a. and make   4. General criticism (yellow).
the yellow cd. for B   5. Partial collections (white).

       6. Individual works (white) alphabeted by titles and each followed
       by: — 7. Criticism of that work (yellow).
       8. Joint author (white).
       9. Added entry cards, ed. tr. etc. (white).

COLORED CARDS IN SUBJECT CATALOG.

   e. BLUE cards for BIBLIOGRAPHY.

       1. "BEST BOOK" cards to show most desirable treatises or articles,

1 Scf. For fullness of author's name follow rule for subject headings, see 2b.

       both popular and exhaustive. S. C. 45.[1] Arrange at beginning
       of subject, preceded only by special topic guide, if any.

       2. "SUBJECT BIBLIOGRAPHY" cards. These take the place of white

2 no2 Or let these duplicate white subject cards in 016 and its subdivisions. Enter, e. g. a bibliography of mathematics on a white card and arrange in 016.51; also on a blue card and arrange in 510.

       subject cards in 016 and its subdivisions. Enter e. g. a bibliog-
       raphy of mathematics on a blue card in black ink, in form of an
       ordinary subject card and arrange after "best book" card, if any,
       in 510.[2]

3 M. For fullness of biographee follow rule for subject headings, see 2b; give only surname of author.

   f. GREEN cards for BIOGRAPHY. S. C. 42.[3] These are practically "best
       book" cards on biography illustrative of the subject. Enter on
       green cards *only the best* lives, as full lists can be found under
       920. Arrange after blue bibliographical cards and before white
       cards.

... make yellow
for subj. cat. with
making one for author
course not true
book plate has
no and no. make the
two cards.

   g. YELLOW cards for CRITICISM. 1. General criticism of an author,
       mainly confined to literature. Make subject card similar to
       S. C. 44.
       2. Criticism of an individual work. S. C. 43. Arrange next after
       works criticised.

rit. of being (1 yellow)
d. without imp.
on auth. cat.   h. ARRANGE cards in subject catalog: —
       1. Special topic guide (white).
       2. Best book card (blue).
rit. of life of Christ   3. Subject bibliography cards (blue).
= yellow cds + only   4. Best lives card (green).
their one cancel cd.   5. General works (white).
       6. Special topics (white), alphabeted by words on top line.
       Criticism cards (yellow) are arranged next after works criticised.

20

9. — MISCELLANEOUS.

ʳ a. In *adding* another entry to a card, — indicates omission of heading — — of heading and title. Add later to earlier EDITIONS in this way, giving full imprint of second and part of title if it differs from the earlier edition. S. C. 32 and 33.

ᵗ b. Use ARABIC FIGURES for all numbers, but in the title follow the form given after names of rulers and popes. S. C. 26. Thus, use figures for numeral adjectives and nouns in any language ; but not for numeral *adverbs* e. g. "printed for the 1st time," but "now first printed."

ᵗ c. Confine author, and subject entry if possible, without omitting important information, to ONE CARD each.

ᵗ d. When more space is needed, number and tie on "EXTENSION CARDS" and indent them as if all on one deep card, leaving space for call number in the upper left corner of author cards. On series cards only, give an abbreviated heading on the back to identify if misplaced.

> *Book numbers.* The letters (or letters and figures) on the second line of the left upper corner of the sample cards are the book numbers as used at Columbia. This determines order on the shelves, but does not affect the card catalog rules. For explanation see shelf list rules.

If preferred titles (of periodicals, anonymous and other books widely known by their title) may be scattered through the name catalog instead of being kept in a separate alphabet. No change whatever is required in the cards. Cards can be found quicker if names and titles are kept separate, as on an average only half as many cards occur under any word, but careless readers may look in the wrong alphabet or forget that there are two. If they are kept separate, guard against mistakes by very plain labels on each catalog.

The subject catalog is to be used with the printed index. Readers need not know anything of its arrangement. This Relativ Index has some 20,000 topics plainly printed for the easiest possible reference. Each is followed by a simple arabic number. Reference to the subject catalog, arranged in simple numerical order, shows the library's resources in that topic. This reference is found by long trial to be much quicker than to the common catalog arranged alphabetically by topics. This enables us to combine the great advantages of classed and dictionary catalogs.

For full explanations see preface to " Decimal Classification and Relativ Index by Melvil Dewey," publisht by Library Bureau, 146 Franklin St., Boston.

SAMPLE CARDS ILLUSTRATING THE CATALOG RULES.

The italics in S. C. 4, 5, 28, 29, 30 and 32 indicate that the matter thus printed is to be written in pencil on the cards.

Numbers in hair line type above and below the regular call number are reference numbers, and should be stamped or written in red ink. S. C. 27, 42, 43, 45, 47-51.

Names and titles in hair-line type at the top of colored cards are written in red ink. S. C. 26, 27, 38, 40, 41, 43, 44 and 46.

✓1. Simple subject card. Rules 1a, 2—6,9.

330.4
M59   Mill, J: S.   *dates*

Essays on some unsettled questions of political economy.   Ed.3.   6+₍1₎+164p.O.   L. 1877.

O

✓2. Simple author card. Rules 1a, 2—6,9.

330.4
M59   Mill, J: Stuart,

Essays on some unsettled questions of political economy.   Ed.3.   6+₍1₎+164p. O.   L. 1877.

O

3. Order of imprint. Imaginary book. Rule 4a.

Smith, John,   *dates.*

Works.   Ed.2 enl.   340+₍1₎p. il. (gr. of por.) (por. of gr. 1 por. pl. phot. 7maps, fac-sim. tab.   O. n.t-p.   N.Y. 1879, '75₎.

22

✓4. Main entry under editor. Rules 1b,1h.

973
W73   Winsor, Justin,   ed.
             Narrative & critical history of America.   *v.2-4,*
il. pl. maps, Q.   B.[1886].

✓5. Official body made author. Rules 1c,1f,1m,1n,2f.

027.073 U. S.—Interior, Dept. of—Education, Bureau of.
U   Public libraries in the U. S... their history, condition & management. Special report.   *pt.1,*O.   W.1876.

✓6. Subject card with editor. Rules 1g,1k,1m,1t,1y,1z,4m,4q.

875.1   Cicero, M. T.
J2   ✓ Opera omnia qvae exstant; a D. Lambino ...
emendata & avcta ... eiusdem D. Lambini annotationes.
2v. in 1, F.   P. in aedibus Rouillij, 1565–66,[v.1,'66].
v.2 apud Bernardum Turrisanum.

✓7. Author card with editor. Rules 1g,1k,1m,1t,1y,1z,4m,4q.

875.1   Cicero, M[arcus] Tullius.
J2   ✓ Opera omnia qvae exstant; a D. Lambino ...
emendata & avcta ... eiusdem D. Lambini annotationes.   2v.
in 1, F.   P. in aedibus Rouillij, 1565–66,[v.1,'66].
v. 2 apud Bernardum Turrisanum.

✓8. Editor card. Rules 1g.1k,1m,1t,1y,1z,4m,9a.

875.1   ✓ Lambin, Denis,   ed. & comnt.
J2   Cicero, M. T.
             ✓ Opera omnia. 1565–66.

✓9. Partial translator card. Rule 1k.

872.3   Warner, Ric... tr.
K   Plautus, T. M.
v. 3-5   Comedies. 1769–74. v. 3–5.

23

10. Author card with title entry. Rules 1u,1v,12,2g.

370.1
R76   Rousseau, J: J_acques_.   dates.
1780.   Émile; ou, De l'éducation.   2v.sq.Q.   Genève

11. Title card. Rules 1u,1v,12,2g.4m,9a.

370.1   Émile; ou, De l'éducation.   1780.
R76   Rousseau, J! J.

12. Partial title card. Rules 12,4m.

P817.39   Biglow papers.
B
            Lowell, J. R.
            Meliboeus-Hipponax; the Biglow papers.   1848–67.

13. Joint author card. Rules 1k,1l,4p,5p,4q.

832.62   Schiller, _J; Christoph F; von_ & Göthe, _J;
U   W. von_.
                Briefwechsel in...1794–1805.   6v. in 3, S.
Stut. 1828–29.
                v.4 has t.-p.w.

14. Joint author card. Rules 1k,1l,4p,5p.

832.62   Göthe, _J; Wolfgang von_ & Schiller, _J; C.
U   F; von_.
                Briefwechsel.   1828–29.

15. Anonymous title card. Author not found. Rules 1f,3k.

823.89
W58   Whitecross & the bench; a reminiscence of the past,
by the author of Five years penal servitude...   10+282p.D.
L, 1879.
                            24

16. Anonymous author card. Author found. Rules 1f,2g,3k,4q.

P560 A
eml.

Chambers, Rob. anon.

Vestiges of the natural history of creation. Ed. 10

12+325+67p.il.O. L. 1853.

See Athenæum 1884, 83:535.

17. Anonymous title card. Author found. Rules 1f,2g,3k.

P560 A

Vestiges of the natural history of creation. 1853.

Chambers, Rob.

18. Author's initials card. Author not found. Rule 1d.

811.49 Ot4

O, G. E.

Thurid & other poems. 123p.O. B. 1874.

19. Author's initials. Reference card. Rule 1d.

O, G. E. see

Otis, G: Edm

20. Pseudonym author card. Real name unknown. Rule 1e.

320.1 C83

Coverdale, Sir H: Standish, pseud.

Fall of the great republic, 1886-88. 226p. S.

N.Y.1895, B.1885.

21. Pseudonym author card. Real name found. Rule 1e.

P823.82 Z2
sq.O.

Thackeray, W: Makepeace.

Mrs Perkins's ball, by M. A. Titmarsh. 46p.il.pl.

L. n.d.

22. Pseudonymous book. Real name found. Added entry card. Rule 1e.

P823.82 Z2

Titmarsh, Michael Angelo, pseud. of

Thackeray, W: M.

Mrs Perkins's ball. n.d.

25

Title card always for pseud. book

23. Several books under same pseudonym. Reference card. Rule 1e.

Eliot, G: pseud.   see
Cross, Mrs M.. Ann (Evans) Lewes.

24. Author analytical "in" card. Rules 1x,4r.

**330.8
Sch5
v.5**

Hasbach, W; c\m ,

Das englische arbeiterversicherungswesen; geschichte
seiner entwickelung & gesetzgebung.   16+447p.O.   Lpz.
1883.   (in Schmoller, Gst.. Staats & socialwissenschaftliche
forschungen. 1883. v.5, pt.1.)

25. Author analytical "see" card. Rules 1x,4r. cf. S.C. 51.

**827.7
D49
v.6**

Milton, J: n,   dates.
Mask of Comus ... (see Deverell, Rob, Discoveries
in hieroglyphics. 1813. v.6, p.81-230.)

26. Biographical subject card. Rules 1w,5n,7f.

**P923.144
A**

Louis 14,   king of France,   1638-1715.
James, G: P. R.
Life & times of Louis the fourteenth.   New ed.
2v. 2por. D.   L. 1851.

27. Biographical analytical subject card. Rules 1y,5n,7f,i.

928.33
**833.62
0
v.l**

Richter, Jean Paul Fr.es- cn,   1763-1825.
Carlyle, T:homas,   dat:..
Biographical sketch of ... Richter.(see Richter. J:P. F;
Flower, fruit & thorn pieces. 1863. v.1, p.1-67.)

28. Periodical subject card. Rules 1f,1t,1u,1z,3j,3k,4f,4q.7e.

**051
N**

Nation; a weekly journal devoted to politics,
literature, science & art, 1865-*date.*   v.1-*date.* sq.F.
N.Y. 1865-*date.*

v.34-37. 1882-83 wanting.

29. Periodical title card. Rules 1f, 1t, 1u, 1z, 3j, 3k, 4f, 4q, 7e.

051
N   Nation; a weekly journal devoted to politics, literature, science & art, 1865–*date.*   *v.1–date.* sq.F.

N.Y. 1865–*date.*

v.34-37, 1882-83 wanting.

30. Series card. Rules 1f, 1v, 2k, 4s, 4t, 5p.
| 1st of 2 cards.

Van Nostrand, D: ed.

Science ser.

628.8
N3 v.5 Butler, W. F. Ventilation of buildings. 1873.

622.41
N5 v.14 Atkinson, J. J. Friction of air in mines. 1875.
                                        See next card.

621.5
N8 v.40 Zahner, R. Transmission of power by compressed
              air. 1878.

621.5
N9 v.46 Ledoux, C: Ice-making machines. 1879.

31. Author card with series note. Rules 1f, 4s, 5p.

621.5
N9 Ledoux, C:
              Ice-making machines, theory of. their action ... tr.
fr. the French...   150p. il.S.   N.Y. 1879.   (Van Nostrand, D: ed. Science ser. v.46.)

32. Author card showing added edition. Rule 4q, 9a.

888.5
Q6 Aristoteles,
              Ethics; Greek il, with essays & notes by Sir A.
Q7 Grant, bart.   Ed.2.   2v. O.   L. 1866.
              ———— ———— Ed.3.   2v.O.   L. 1874.
v.2 of ed. 3 wanting.

27

33. Editor card showing added edition. Rule 9a.

888.5
Q6   Grant, Sir Alex: bart.   ed.

     Aristoteles.

       Ethics. 1866.

Q7   —— ——1874.

34. Title card for sacred books. Rule 11.

P297
B   Koran.

       Koran ... tr. into English ... by G: Sale.   New ed.
with a memoir of the translator.   16+516p. pl. maps, O.
L. 1850.

35. Reference card. Rule 2a.

     Groot, Hugo van,   see

     Grotius, Hugo.

36. Reference card. Rule 10.

     Sidney, H: see

     Romney, H: Sidney, 1st earl of.

37. Title analytical card. Rule 12.

822
In2   Julius Cæsar; a tragedy. 1824.
v.6   Shakspere, W: (in Inchbald, Mrs E. (S.) British
theatre. 1824. v. 6.)

38. Biographical subject card. Rules 1w.5n,7f.

923.144   *Napoleon 1, emperor of France,*   1769–1821.
N1614   Jomini, A. H: baron de.

       Life of Napoleon... tr. fr. the French with notes by
H: W. Halleck.   4 v. O & Atlas obl. O.   N. Y. 1864.

39. Author card for biography. Rule 1w.

923.144
N1614 Jomini, Ant. H: baron de,   dates

Life of Napoleon... tr. fr. the French with notes by H: W. Halleck.   4 v. O & Atlas obl. O.   N.Y. 1864.

40. Green card for biography in author catalog. Rule 8b.

For biography of

    Napoléon 1, emperor of France, 1769-1821. see

923.144
N1614 Jomini, A. H: baron de,   dates.

    Life of Napoleon... tr. fr. the French with notes by H: W. Halleck.   4 v. O & Atlas obl. O.   N.Y. 1864.

41. Green card for autobiography in author catalog. Rule 8b.

For biography of

    Carlyle, Thomas,   em   *1795-1881,* see

928.24
C191 Carlyle, T: 1795-1881.

    Reminiscences; ed. by J. A. Froude.   10+[2]+536p. O.

N. Y. 1881.

42. Green card for biography in subject catalog.

370.9 For valuable matter bearing on this subject see
923.7 (blue)

    Biography of **education.**

Among other lives note specially: —

Call number.		Life of	By
923.74	Ar62	Arnold, T: Thomas.	Stanley.
923.743	P43	Pestalozzi, J: H:	Biber.
923.77	M31	Mann, Horace.	Mann, Mrs
923.77	L992	Lyon, Mary.	Fiske.

43. Yellow card for criticism in subject catalog. Rule 8g.

For criticism of

215   *Seeley, J: Rob.*   dates

    Natural religion.   em   see

824.89
M99 Myers, F: W. H.   dates.

    New Eirenicon. (see Myers, F: W. H. Essays, modern.

1883. p. 289-311.)

_____ 44. Yellow card for criticism in author catalog.   Rule 8c. _____

For criticism of

    Browning, Robert   dates

                           see

821.83 Nettleship, J: T.   dates
D
    Essays on ... Browning's poetry.   6 + [1] + 305p.D.
oni.
L. 1868.

—

'45. Best book card in subject catalog.   Rule 8e. _____

335
                   For popular treatises see : —

335 E19 Ely, R: T.   Fr. & Ger. socialism.
335 L381 Laveleye, E.L.v.&e Socialism of to-day.
335 R12 Rae, John Contemporary socialism.
335 W88 Woolsey, T. D.   Communism & socialism.

   For scholarly or exhaustive treatises see : —

335 M57 Meyer, R.   Emancipations-kampf.
335 M73 Molinari, Gustave Mouvement socialiste.
335 R33 Reybaud, L: Socialistes modernes.
335 Sch1 Schäffle, A. E.F. Kapitalismus.
Revised *A. H. B. My 89.*

—

'46. Blue card for bibliography in author catalog.   Rule 8a. _____

For bibliography of

    Carlyle, Thomas   dates   i cm   see

012 [Shepherd, R: H.]   dates anon.   Comp.
C
    Bibliography of Carlyle ; a ... list ... in chronological
order of [his] published writings ... 1820-81.   11+60p. D.
L. [pref. 1881].

✓47. Special topic guide. Rule 7m.

Berkshire co. Mass.

974.41.   Cards on this subject stand in this order :—
Bibliographical (blue cards); Biographical (green); General works (white);
Special topics (white); alphabeted by words on top line, in red ink, as follows :—

Great Barrington.

Lee.

Lenox.

Pittsfield.

Stockbridge.

Williamstown.

✓48. Main subject card showing general subject reference. Rule 7h.

813.49   Wallace, Lew[is].   *dates*
W15   Fair god ; or, Last of the 'Tzins, a tale of the con-
quest of Mexico...   Ed.25.   14+586p. O.   N.Y.
972.014   1886, ['73].

✓49. General reference subject card. Rule 7h.

972.014   Wallace, Lew[is],   *dates.*
813.49   Fair god ; or, Last of the 'Tzins, a tale of the con-
W15   quest of Mexico. 1886.

✓50. Main subject card showing analytical subject references. Rule 7i.

917.8   Ludlow, F. H.   *dates*
L96   Heart of the continent; a record of travel across
the plains & in Oregon, with an examination of the
917.9   Mormon principle.   6+568p. pl. O.   N.Y. 1870.
298

✓51. Analytical reference subject card. Rule 7i.   *See S.C.24 for auth. analytical "see" card*

298   Ludlow, F. H.   *dates.*
917.8   Utah's life principle & destiny. (see his Heart of
L96   the continent. 1870. p. 503–568.)

'52. Cards showing contents.
~~1st of 7 cards.~~

813.33 Hawthorne, Nathaniel.

    Complete works; with introd. notes by G: P. Lathrop,
& il. with etchings.   12v. 1por. 11pl. O.   B. 1886, ['50-83].
(Riverside ed.)

Contents.

    v.9 Amer. note-books.

    v.5 Blithedale romance.   **See next card.**

2

813.33   v.11 Dolliver romance.

    v.7-8 English note-books. 2v.

    v.11 Fanshawe.

    v.10 French & Italian note-books.

    v.4 Grandfather's chair.

    v.3 House of the 7 gables.

    v.6 Marble faun.   **See next card.**

The words "Contents," "~~1st of~~ cards," and "see next card," S. C. 30 and 49, are stamped on the cards in blue ink, to secure greater distinctness.

SAMPLE TITLE PAGES ILLUSTRATING CATALOG CHECK MARKS.

MRS. PERKINS'S BALL.

BY M. A. TITMARSH

W: Makepeace Thackeray

CHAPMAN & HALL, 186 STRAND.

L.

n. d.

NOTE ON FIRST TITLE. The cataloger supplies in hard pencil on the title page lines 3, 5 and 6. Also the "ichael" and "ngelo" written at an angle for lack of room between initials. Line 3 gives the author's real name. L is the regular sign for London, ~~and n. d. for no date of publication.~~

FLOWER, FRUIT, AND THORN PIECES

OR THE

MARRIED LIFE, DEATH, AND WEDDING

OF

THE ADVOCATE OF THE POOR,,

Firmian Stanislaus Siebenkäs.

BY

JEAN PAUL FRIEDRICH RICHTER.

Translated from the German
By EDWARD HENRY NOEL,
With a Memoir of the Author,
By THOMAS CARLYLE.

IN TWO VOLUMES.

I.

BOSTON:
TICKNOR AND FIELDS.
1863.

Library abbreviations

Compiled by MELVIL DEWEY.

100 FORENAMES. CUTTER ABBREVIATIONS.

Ab.	Abraham	F.. s.	Frances
Alex.	Alexander, Alexandre	F:	Frederick, Friedrich,'
Alf.	Alfred		Frédéric
And.	Andrew, Andreas, André	G:	George, Georg, Georges
A..	Anna	Gert.	Gertrude, Gertraud
Ant.	Anthony, Anton, Antoine	Gilb.	Gilbert
		Gi. Bat.	Giovanni (Giam) Battista
Arch.	Archibald, Archambaud	G..	Grace
Art.	Arthur	Greg.	Gregory, Gregor, Gregoire
A:	Augustus, August, Auguste	Gu.	Guillaume, Gulielmus
A: a.	Augusta	Gst.	Gustavus, Gustav, Gustave
A: in.	Augustin		
A: inus.	Augustinus	H..	Helen
Bart.	Bartholomew, Bartholomäus, Barthélemi	H:	Henry, Heinrich, Henri
B..	Beatrice	Hrm.	Herman, Hermann
B:	Benjamin	Hip.	Hippolyte, Hippolytus
Bern.	Bernard, Bernhard	Hu.	Hugh, Hugo, Hugues
Cath.	Catherine, Catharine	Ign.	Ignatius, Ignaz, Ignace
C:	Charles, Carl		
C..	Charlotte	I:	Isaac, Isaak
Chris.	Christopher, Christoph (f), Christophe	I..	Isabella
		Jac.	Jacob, Jacques
Clar.	Clarence	Ja.	James
Dan.	Daniel	J..	Jane
D:	David	J:	John, Johann, Jean
D..	Delia	Jos.	Joseph
Edg.	Edgar	Jose.	Josephine, Joséphe
Edm.	Edmund, Edmond	Jul.	Julius, Jules
E:	Edward, Eduard, Edouard	K:	Karl
		K..	Kate
E..	Elizabeth	Kath.	Katherine
Ern.	Ernest, Ernst	Lr.	Lawrence, Laurence, Lorenz, Laurent
Eug.	Eugene, Eugen		
F..	Fanny	L:	Lewis, Ludwig, Louis
Fer.	Ferdinand	L..	Louisa
Fitz W:	Fitz William	L: e.	Louise

Library abbreviations

Marg.	Margaret, Margarethe	**Rob.**	Robert
	Marguerite	**S:**	Samuel
M:	Mark, Marcus, Marc	**S..**	Sarah
M..	Mary	**Seb.**	Sebastian, Sébastien
Mat.	Matthew, Mathäus,	**Ste.**	Stephen, Stephan
	Mathieu	**Thdr.**	Theodore, Theodor
N..	Nancy	**T..**	Theresa
N:	Nicholas, Nikolaus,	**T:**	Thomas
	Nicolas	**Tim.**	Timothy, Timotheus,
Ol.	Oliver, Olivier		Timothée
O..	Olivia	**U:**	Ulrich
O:	Otto	**U..**	Ursula
Pat.	Patrick	**V:**	Victor, Viktor
P.. a.	Paulina	**V..**	Victoria
P..	Pauline	**Wa.**	Walter, Walther
P:	Peter, Pierre	**Wash.**	Washington
Ph.	Philip, Philipp,	**W..**	Wilhelmina
	Philippe	**W:**	William, Wilhelm
R..	Rebecca	**Zach.**	Zachary
R:	Richard	**Z..**	Zenobia

Where : and .. is used in English names, use ; and ., for the German form, and : and ,. for the French. e.g., J: John, J; Johann, J: Jean.

FOR HEADINGS.

Besides the preceding 100 forenames.

abr.	abridger	**Gt. Br.**	Great Britain
aftw.	afterwards	**pseud.**	pseudonym
annot.	annotator	**pub.**	publisher
anon.	anonymous	**supt**	superintendent
b.	born	**tr.**	translator
col.	collector	**U. S.**	United States
comnt.	commentator	**&**	and
co.	company	( ) include maiden name of married
comp.	compiler	woman.
contin.	continuer	[ ] include words or parts of words
dept.	department	supplied
d.	died	? after a word or figure means *prob-*
ed.	editor	*ably, perhaps.*

Use also the common abbreviations for political, military, professional and honorary titles.

FOR IMPRINTS AND NOTES.

Use the size symbols, F Q O D S T Tt Fc, given at the end.

c	copyright, c. g., 1882 ['80]	**fac-sim.**	fac-similes
cm	centimeter	**gr. of por.**	group of portraits
col.	columns	**il.**	illustrated—ions
ed.	editions	**l.**	leaves
f.	folios		

mut.	mutilated	por. of gr.	portrait of group
n. t-p.	no title-page	pt.	part
p.	page or pages	ser.	series
p.	published, e. g. 1882	tab.	tables
	[P'80]	t-p.	title-page
phot.	photographs	v.	volumes
pl.	plates	v. p.	various paging
por.	portraits	w.	(before words) with
		w.	(after words) wanting

In notes, the abbreviations in all these lists may be used.

FOR BOOK TITLES.

Besides the abbreviations for honorary and other designations.

acct.	account	med.	medical
ad.	additions—al	mem.	memoir
Am. or Amer.	America—n	misc.	miscellaneous
anal.	analysis—tical	ms. mss.	manuscript—s
ap.	appended	N. A.	North America
apx.	appendix	nouv.	nouvelle
biog.	biography—ical	pref.	preface, prefatory
chron.	chronology—ical	pub.	published—rs
comp.	compiled	rel.	relating—ive
cont.	containing, contents	rept.	report—ed—er
contin.	continuation,	rev.	revised—ion
	continued	S. A.	South America
cor.	corrected	sep.	separate
dept.	department	soc.	society
ed.	edited—or—ion	sup.	supplement—ary—ing
Eng.	English	theol.	theology—ian
enl.	enlarged	tr.	translated, traduit,etc.
Fr.	French	trans.	transactions
fr.	from	U.S.	United States
geog.	geography—ical	vocab.	vocabulary
geol.	geology—ical	&	and, in all languages
geom.	geometry—ical	[ ]	words or part of words
hist.	history—ical		supplied
hrsg.	herausgegeben	—	to and including, or
Ger.	German—y		continued
Gr.	Greek—cian	. . .	matter omitted
impr.	improved—ments	?	probably, perhaps
incl.	including	‖	transition to another
introd.	introduction—ory		page
Ital.	Italian	ǀ	end of line on title
Lat.	Latin		page. Used in ex-
lib.	library		act bibliographical
lit.	literature—ry		work.

NEVER use title abbreviations for specially prominent words.

Library abbreviations

FOR PLACES OF PUBLICATION.

Or use fuller form on cards. *Use first on cards.* *In accession and all official records use shorter form.*

Alb.	Albany	Lpz.	Leipzig
Amst.	Amsterdam	Lug. Bat.	Lugduni Batavorum
B. or Bost.	Boston	Mil.	Milano
Balt.	Baltimore	Mün.	München
Ber.	Berlin	N. O.	New Orleans
Brns.	Braunschweig	N. Y.	New York
Camb. or Cb.	Cambridge	Ox.	Oxford
Chic. or Ch.	Chicago	P. or Par.	Paris
Cin.	Cincinnati	Ph. or Phil.	Philadelphia
Copng.	Copenhagen	San Fran. or S. F.	San Francisco
Dub.	Dublin	St. L.	St. Louis
Edin. or Ed.	Edinburgh	St. Pet. or St. P.	St. Petersburg
Eng.	England	Stut.	Stuttgart
Fir.	Firenze	U S.	United States
Glasg. or Gl.	Glasgow	Ven. or V.	Venice
Göt.	Göttingen	W. or Wash.	Washington
Kjöb.	Kjöbenhavn		
L. or Lond.	London		
Ley.	Leyden		

Also the common abbreviations for the states. Use for all languages when the equivalent name contains these letters.

TITLES, STATES, ETC.

A. B.	bachelor of arts	C. S. N.	C. S. navy
abp.	archbishop	Ct.	Connecticut
A. D.	year of our Lord	D. C.	District of Columbia
adjt.	adjutant	D. C. L.	doctor of civil law
adm.	admiral	D. D.	doctor of divinity
Ala.	Alabama	Del.	Delaware
A. M.	master of arts	dist.	district
Am. or Amer.	American	D. T.	Dakota territory
A. R. A.	associate of the royal	Eng.	England
Ariz.	Arizona academy	Fla.	Florida
Ark.	Arkansas	F. R. S.	fellow of the royal
atty.	attorney		society
B. A.	bachelor of arts	Ga.	Georgia
bart.	baronet	gen.	general
B. C.	before Christ	gov.	governor
bp.	bishop	Gt. Br.	Great Britain
brig. gen.	brigadier general	Ia.	Iowa
Cal.	California	Id. T.	Idaho territory
capt.	captain	Ill.	Illinois
card.	cardinal	Ind.	Indiana
Col.	Colorado	Ind. Ter.	Indian territory
C. S. A.	Confederate States of	jr.	junior
	America or C. S.	Kan.	Kansas
	army		

38

Library abbreviations

Ky.	Kentucky	N. Y.	New York
La.	Louisiana	O.	Ohio
L. I.	Long Island	Or.	Oregon
LL. B.	bachelor of laws	Pa.	Pennsylvania
LL. D.	doctor of laws	pres.	president
lt.	lieutenant	R. A.	royal academician
maj.	major	Rev.	reverend
marq.	marquis	R. I.	Rhode Island
Mass.	Massachusetts	R. N.	royal navy
M. A.	master of arts	S. A.	South America
M. C.	member of Congress	S. C.	South Carolina
M. D.	doctor of medicine	sc.	sculpsit, engraver
Md.	Maryland	sen.	senior
Me.	Maine	S. T. D.	doctor of sacred
Messrs.	plural of Mr.		theology
Mich.	Michigan	supt.	superintendent
Minn.	Minnesota	Tenn.	Tennessee
Miss.	Mississippi	Tex.	Texas
Mlle.	mademoiselle	U. S.	United States
Mme.	madame	U. S. A.	U. S. of America or
Mo.	Missouri		U. S. army
M. P.	member of Parliament	U. S. N.	U. S. navy
Mr.	mister	U. T.	Utah territory
Mrs	mistress	Va.	Virginia
N. A.	North America	visc.	viscount
N. B.	New Brunswick	Vt.	Vermont
N. C.	North Carolina	Wis.	Wisconsin
Neb.	Nebraska	W. T.	Washington territory
N. H.	New Hampshire.	W. Va.	West Virginia.
N. J.	New Jersey		
N. M.	New Mexico		
N. S.	Nova Scotia		

MONTHS.

Ja F Mr Ap My Je Jl Ag S O N D

DAYS.

Sn M Tu W Th F St

Use in this order " W 9 S 85" for " Wed. Sept. 9th, 1885."
Do not use abbreviations for days and months on catalog cards.

FIGURES.

Never use roman numerals. Use arabic figures, a half larger than the
script, for all numerical expressions.

39

Library abbreviations

Size Notation.

FOLD SYMBOL.	SIZE LETTER.	OUTSIDE HIGHT.
Never use for size.	*Never use for fold.*	*in centimeters.*
48°	Fe	Up to 10
32°	Tt	10 " 12.5
24°	T	12.5 " 15
16°	S	15 " 17.5
12°	D	17.5 " 20
8°	O	20 " 25
4°	Q	25 " 30
f°	F	30 " 35
	F⁴	35 " 40
	F⁵	40 " 50
	F⁶	50 " 60

For all books over 35cm high the superior figures show in which 10cm of hight the book falls, e. g., F8 is between 70 and 80cm high.

Prefix *nar.* if width is less than ⅔ hight.
" *sq.* " more " ¾ "
" *ob.* " more than hight.
These dividing lines will be remembered by the three threes ⅔, ¾, ⅓.

ACTUAL SIZE METHOD.

Give all sizes in cm (for great accuracy add decimals), leaving the old sysbols and names, 8° and Octavo to indicate fold only. Give hight first, followed by h, or by x and width, e.g., 23ʰ or 23 x 14. 23ʰ means between 22 and 23, i. e., in 23ᵈ cm. All measures are taken outside the cover. Width is from hinge to edge not including the round. To measure paper or letter-press prefix p(aper) or t(ype) to figures, including in type neither folio nor signature lines; e.g., 23x14, p22x14, t17x10, 8° describes a book with size of cover, of paper, of letter-press, and fold.

Library Colors.

The day colors are used, e. g., on Inspection Shelves, to mark by a colored thred the day on which each book is to go to its regular place in the library; or, they mark the day of a loan or mem., if the slips used are changed daily, as in some libraries. They are :—

Sn.	M.	Tu.	W.	Th.	F.	St.
Pink.	Blue.	Green.	White.	Red.	Fawn.	Salmon.

The language colors used, e. g., in bindings are :—

1. American, Light Brown.
2. English, Dark Brown.
3. German, Black.
39. Minor Teutonic, Dark Blue.
4. French, Red.
5. Italian, Maroon.
6. Spanish, Olive.
7. Latin, Light Green.
8. Greek, Dark Green.
91. Minor Aryan, Light Blue.
92. Semitic, Yellow.
93-99. Hamitic, etc., Light Drab.

ABBREVIATIONS FOR BINDINGS.

bds.	boards	dk.	duck	rox.	roxburghe
buck	buckram	mor.	morocco	rus.	russia
cf.	calf	pap.	paper	sh.	sheep
cl.	cloth, muslin	ro.	roan	vel.	vellum

Bibliography of catalog rules.

1750 Francke, J; Michael. "Jam de consilio et dispositione hujus catalogi, de causis item, unde natum omne hoc institutum, breviter exponam." (see his Catalogus bibliothecæ Bunavianæ. 1750–56. v. 1. pref. p. 6–14.)

This preface is valuable for the minute explanation given of the extent to which "analyticals" have been carried in the body of the catalog. "This systematic catalog according to subjects is a master-piece not to be surpassed," says Ebert. c: A. N.

1761 Audiffredi, Giovanni Battista. "Ratio qua indicem conscribendum esse consti-tueramus." (see his Bibliothecæ Casanatensis catalogus, 1761, v. 1, pref. p. 10–18.)

A full explanation of the "rules" followed in making "this unsurpassed model of an alphabetical catalog according to the names of authors." But A. often takes the liberty of changing the wording of titles. c: A. N.

1790 Kayser, Albrecht Christoph. Über die manipulation bey der einrichtung einer bibliothek & der verfertigung der bücherverzeichnisse. 18 + 123 p. D. Bayreuth 1790.

Gives illustrated directions for putting up movable shelves and for numbering and locating books, rules for cataloging and alphabeting, and a form for shelf-list. Puts all books in two classes: where author is known, and anonymous. Discusses double titles and how anonymous books should be entered. c: A. N.

1811 Ebert, F! Adolph. Die erfordernisse des alphabetischen katalogs. (see his Über öffentliche bibliotheken besonders deutsche universitätsbibliotheken. 1811. p. 40–46.)

1829 [Chaillot, P?] Manière de faire le catalogue d'une grande bibliothèque. (see his Manuel du libraire. 1829. p. 17–34.)

1829 Schrettinger, Martin. Versuch eines vollständigen lehrbuchs der bibliothek-wissenschaft... 2 v. D. Mün. 1829.

Rules for subject & subject catalogs with specimen pages. Dziatzko notes an edition in 1808

1834 Namur, J! P! [The catalogue.] (see his Manuel du bibliothécaire. 1834. p. 61–95.)

1834 Schrettinger, Martin. Anleitung zur verfertigung eines alphabetischen namen-kataloges. (see his Handbuch der bibliothek-wissenschaft. 1834. p. 43–85.)

A very elaborately classified hand-book of library economy. c: A. N.

1836 Richter, Benedikt. Von der herstellung der ordnung. (see his Kurze anleitung eine bibliothek zu ordnen & in der ordnung zu erhalten. 1836. p. 2–38.)

Gives 14 numbered rules; discusses these, and other cataloging problems in 20 sections; folded plates of forms. c: A. N.

1839 Hesse, Leopold A! Constantin. (L. A. Constantin, pseud.) Des catalogues. (see his Bibliothéconomie. 1839. p. 95–130.)

New ed. 1841; Ger. tr. Lpz. 1840.

1840 Hesse, see 1839.

1841 British museum. Rules. (see its Catalogue of printed books. 1841. v. 1, pref. p. 5–9.)

The famous 91 rules compiled by Panizzi, T: Watts, J. Winter Jones, J. H. Parry & E: Edwards. Adopted by the trustees, 13 Jl. 1839.

—— (see Nichols, T: Handbook for readers at the British museum. 1866. p. 38–54.)

Nichols arranges the subjects treated of in the rules alphabetically & does not always preserve the exact wording.

—— Règles à suivre pour la confection du catalogue du Musée Britannique; [tra-duites par Olivier Barbier.] (see Bulletin du bibliophile. 1845. p. 299–308, 338–347.)

—— [Rules tr. into Portuguese.] (see Ramiz-Galvao, B: F. Bibliothecas publicas de Europa. 1874.)

—— Such portions of [these rules] as appear likely to be of benefit to readers in cases of more than ordinary difficulty. (see Sims, R: Handbook to the library of the British museum. 1854. p. 91–95.)

Extracts from 18 rules are given.

—— Rules for the compilation of the catalogue, as revised & codified to Jl. 1862. (see Stevens, H: Catalogue of American books in the British museum. 1866.)

1841 Hesse, see 1839.

1844 **Vituoni, Francesco.** Sul modo di compilare il catalogo di una biblioteca; saggio proposto al giudizio dei dotti. 24 + 302 p. O. Mil. 1844.

1845 **British museum, see 1841.**

1848 **Hébert, J. B.** Essai sur la formation d'un catalogue général des livres & manuscrits existant en France à l'aide de l'immatriculation . 11 + 128 p. O. P. 1848.
Of no practical value at the present day, but of some interest as a very elaborate scheme by a man who thought that the salvation of France was to be brought about by cataloging its population, its real property, its public debt, &c. W: C. L.

1850 **Brunet, P! Gustave.** Quelques mots au sujet des difficultés que présente la cata-logographie. (see Bulletin du bibliophile. 1850. p. 563-567.)

1852 **Jewett, C: Coffin.** Smithsonian report on the construction of catalogues of libraries... & their publication by means of separate, stereotyped titles; with rules & examples. 6 + 78p. O. W. 1852.
For preliminary plan of this work, see Amer. assoc. for the advancement of science. 1851. 4: 165-176.

—— —— Ed. 2. 12 + 96 p. O. W. 1853.
See also 1888, Italian translation by Guido Biagi.

1852 **Schleiermacher, Andreas A; Ernst.** Von den bibliothekscatalogen. (see his Bibliographisches system der gesammten wissenschaftskunde. 1852. v. 1, p. 27-63.)

1853 **Jewett, see 1852.**

1854 **British museum, see 1841.**

1855 **Manchester (Eng.) Public free libraries.** Special report on the plan, preparation, & printing of a classed catalogue of the reference department. 22 p. O. Manchester 1855.
Gives specimen pages, but no systematic set of rules. W: C. L.

1855 **Règles suivies pour la confection du catalogue** & dont la connaissance pourra rendre plus facile l'usage de ce travail. (see Bibliothèque impériale—Département des imprimés. Catalogue de l'histoire de France. 1855. v. 1, pref. p. 23-24.)

1855 **Seizinger, J; G;** Bibliothekstechnik; mit einem beitrag zum archivswesen, nebst 44 formularen. [4] + 102 p. + 45 f. O. Lpz. 1855.
Reprinted 1860. P. 20-54 give detailed instructions for author, classed, & subject catalogs, p. 60-94 for alphabetical arrangement of titles. " Formularen " are 44 folios of sample titles.

1856 [Crestadoro, Andrea.] Art of making catalogues of libraries. 60 p. O. L. 1856.

1856 **Petzholdt, Julius.** Von der verzeichnung & aufstellung der bibliothek. (see his Katechismus der bibliothekenlehre. 1856. p. 63-129.)
Contains specimen pages of catalog. Ed. 2, 1871, ed. 3, 1877.

1859 **Edwards, E:** Classification & catalogues; difficulties, rules & details. (see his Memoirs of libraries. 1859. v.2, p. 832-850.)

1859 **Sobolstchikoff, B.** De la manière de dresser les catalogues. (see his Principes pour l'organisation & la conservation des grandes bibliothèques. 1859. p. 47-65.)
Gives 27 rules for cataloging & 4 for alphabeting. They agree in the main with the A. L. A. rules. Advocates fixed location for shelving books; is admirably condensed & contains much valuable instruction. C: A. N.

1863 **Harvard college library.** Illustrations of the new catalogues, [by Ezra Abbott]. (see its Report of the committee of the overseers for 1863. p. 69-76.)
Consists of specimen cards.

1866 **British museum, see 1841.**

1869 **Cutter, C: Ammi.** The new catalogue of Harvard college library. (see North Amer. review. 1869. 108: 96-129.)
Discusses difficulties of author entries (periodicals, societies, &c.): of classed & dictionary catalogs & shows how Ezra Abbot's plan, still used at Harvard, combines alphabeting & classification.

1869 **Perkins, see 1884.**

1870 **Boston—Public library.** Points to be considered in cataloguing, revising & proof-reading. 8 p. Q. n. t.-p. [B.] 1870.
Only 200 printed & long ago out of print.

1870 **Valentinelli, Giuseppe.** Cataloghi alfabetici di libri per nomi di autori. (see Giornale delle biblioteche. 1870. v. 4, no. 4.)

1871 **Petzholdt, see 1856.**

1872 Zenti, Ignazio. Elementi di bibliografia, ossia regole per la compilazione del catalogo alfabetico di una pubblica biblioteca. 11 + 58 + [1] p. S. [Verona] 1872.

Of some value because it discusses the entry of Italian & mediæval names. W: C. L.

1874 [Barack, K; A;] Regeln für die verzeichnung der bücher. A. Zettelcatalog. 13 p. O. Strassburg 1874.

An unusually interesting & in many respects original set of rules for the use of the university library in Strasburg. Brief, but clear & direct. W: C. L.

1874 Biadego, Gius. Dei cataloghi di una pubblica biblioteca e in particolare del catalogo reale. 31 p. O. Verona 1874.

1874 British museum, see 1841.

1876 Cutter, C: Ammi. Rules for a printed dictionary catalogue. 89 p. O. (in U. S.— Interior, Dept. of — Education, Bureau of. Public libraries in the U. S. 1876. pt. 2.)

Indispensable in making dictionary catalogs; the standard authority, often spoken of as the "cataloger's bible." *New edition about Dec. '89 can be had free on application to U. S. Bureau of education, Washington.*

Italian translation in preparation by G. Biagi for Bib. di bibliog. & paleog.

1877 Harvard college library. Rules for the use of capitals, [compiled by J: Fiske] 1 p. O. [Camb. 1877?]

1877 Petzholdt, see 1856.

1877 Hjaltalín, Jón Andrjesson. Remarks on rules for an alphabetical catalogue. (see Conference of librarians in London. Transactions. 1877. p. 93–96.)

1878 —— —— 23 p. T. L. 1878.

1879 Amenduni, Giuseppe. Dell'ufficio del bibliotecario; appunti. O. Nap. 1879.

1879 Cambridge (Eng.) university—Library. Rules to be observed in forming the alphabetical catalogue of printed books. (see Cambridge university reporter. 1879. p. 768–771.)

49 rules.

1879 Dimmock, G: Directions for preparing the bibliographical record for publication in Psyche, organ of the Cambridge entomological club, by the editors. [3]+21 p. T. Camb. [Mass.] 1879.

1880 Bailey, Ja. Blake. Some points to be considered in preparing catalogues of transactions & periodicals. (see Monthly notes. 1880. 1 : 12–16.)

Noted in L. j. 5: 44. Read at monthly meeting of L. A. U. K. 6 F 1880.

1880 [Thomas, Ralph.] [Cataloguing.] (see his Aggravating ladies. 1880. p. 9–39.)

Contents; How to describe a book; — Cataloguing; — Different descriptions of books; — Matters to be attended to in cataloguing; — Style of printing; — Punctuation; — The beginning & the end; — Errors; — Means of identifying the authors of anon. & pseud. publications.
Sensible & helpful but not systematic or extensive enough for independent use. W: C. L.

1881 Firenze — Biblioteca nazionale centrale. Istruzione per la compilazione e la copia del catalogo alfabetico [by Giuseppe Fumagalli.] (see its Regolamento per il servizio. 1881. p. 23–28.)

Not on the market.

1881 Library assoc. of the United Kingdom. Cataloguing rules. (see Transactions. 1881. p. 83–84.)

—— —— (see Monthly notes. 1881. 2:81–84.)

—— —— (see Library journal. 1881. 6:315–316.)

—— —— [As revised at Liverpool, 1883.] (see Library chronicle. 1885. 2: 25–28.)

—— Règles pour la rédaction des catalogues. (see Bibliothèque de l'école des chartes. 1881. 42 : 601–605.)

—— Katalogisierungs-regeln. (see Neuer anzeiger für bibliographie & bibliothek-wissenschaft. 1885. 46:166–172.)

Preliminary reports of the L.A.U.K. committee, also discussion, changes, &c. Lib. assoc. of the U. K. Transactions. 1879. p. 8–10; 1880. p. 5–8, 135–136, 174–178; L. j. 4: 416–17, 5: 271–73.
Notes on L.A.U.K. rules by a German librarian, L. j. 1884. 9:190–191.
In L.A.U.K. Trans. 1881, in Neuer anzeiger & in Library chronicle, 1885, 48 rules are given; in Monthly notes, L. j. and in the French translation, 49 rules. Rule no. 19 of the former is separated into 19 & 20 in the latter.

1882 Bodleian library. Compendious cataloguing-rules for the author catalogue. Broadside. 35.5 x 55^em n. p. 1882.

— — (see Monthly notes. 1883. 4:5–9, 31–33.)

— — (see Library journal. 1883. 8:298–301.) New edition, 1885.

1882 Cousin, Jules. Catalogue alphabétique-méthodique. (see his De l'organisation & de l'administration des bibliothèques. 1882. p. 37–97.)

1882 Milchsack, Gustav. Wie soll man inkunabeln verzeichnen? (see Neuer anzeiger für bibliographie und bibliothekswissenschaft. 1882. p. 15–25 & 49–53.)
Suggests rules to be followed, & gives about 20 titles as examples. G. W. H.

1883 Amer. library assoc. Condensed rules for an author & title catalog. (see Library journal. 1883. 8:251–254. 263–264.)

— Same printed separately.
L. j. 1878. 3:12–20 gives report of the committee on uniform title entries, containing condensed rules with the committee's reasons & the sub-report on sizes.

1883 Bodleian library, see 1882.

1883 Grassauer, Fer. Beschreibung des bibliotheksbestandes. (see his Handbuch für österreichische universitäts und studien-bibliotheken. 1883. p. 86–124.)
Catalog rules for the Austrian state libraries. W: C. L.

1883 Rettig, G; Katalog. (see his Leitfaden der bibliothekverwaltung, hauptsächlich für jugend-und volks-bibliotheken. 1883. p. 21–37.)
Very elementary & based mainly on Petzholdt's Katechismus. G. W. H.

1883 Robert, Ulysse. Instruction pour procéder à la confection du catalogue de chacune des bibliothèques sur lesquelles les directoires ont dû ou doivent incessamment apposer les scellés. (see his Recueil de lois décrets... circulaires... concernant les bibliothèques publiques... universitaires, &c... 1883. p. 11–19.)

1884 Blackburn, C: F. Hints on catalogue titles, & on index entries, with a rough vocabulary of terms and abbreviations... 10+[1]+181 p. Q. L. 1884.
Reviewed in Library chronicle. 1884. 1:78.
" " Bibliographer. 1884. 6:18.
A gossipy, entertaining book, well worth reading, but not to be implicitly trusted. A. N. B.

1884 Laschitzer, Simon. Wie soll 'man kupferstich-& holzschnittkataloge verfassen? (see Mittheilungen des Instituts für österreichische geschichtsforschung. 1884. 4: 565–617.)

1884 Note sur la rédaction des catalogues de manuscrits. 20 p. O. P. 1884.
4 p. of text & 50 sample titles, published by a government commission; see Centralblatt für bibliothekswesen. 1887. 4: 447.

1884 Perkins, F: Beecher. San Francisco cataloguing for public libraries; a manual of the system used in the San Francisco free public library. 53 p. O. San Fran. 1884.
He published a modification of Jewett's rules in Amer. publisher, 1869. Mr Perkins was many years in the Boston Pub. Lib. & is one of the most experienced library and literary workers. These rules with the accompanying "rational classification" are very suggestive & one of the first publications to be read by every student of cataloging. The Library Bureau now supplies the two volumes for $1.00. M. D.

1885 Christiania (Norway) — Universitet. Regler for katalogisering af Norsk literatur. (see its Year-book. 1884. p. 8–10.)

1885 Keysser, Adolf. Über die einrichtung der alphabetischen hauptkataloge öffentlicher bibliotheken. (see Centralblatt für bibliothekswesen. 1885. 2: 1–19.)
Criticised by Mecklenburg, p. 91–96. One of his points defended against M. by Grienberger, p. 327–328. Supplemented by Steiff, see 1885.

1885 Library assoc. of the United Kingdom, see 1881.

1885 Mecklenburg, H. B. Über alphabetishe anordnung. (see Centralblatt für bibliothekswesen. 1885. 2: 345–382.)
P. 345–346 urge importance of catalog rules & give a brief bibliography.

1885 Meier, Gabriel. Wie sollen handschriftenkataloge beschaffen sein? (see Central blatt für bibliothekswesen. 1885. 2: 463–471.)

1885 Ottino, Giuseppe. La biblioteca e il catalogo. (see his Manuale di bibligrafia. 1885. p. 83–146.)
Brief rules for cataloging. W: C. L.

1885 Steiff, K; Über die einrichtung der alphabetischen haupt-kataloge öffentlicher bibliotheken. (see Centralblatt fur bibliothekswesen. 1885. 2: 173-181.)

Supplementary to Keysser's rules.

1886 **Condensed** rules for a card catalog. (see Library notes. 1886. 1:111-131.)

A. L. A. rules enlarged, slightly altered, & illustrated by 36 fac simile sample cards. See also 1888 and 1889. Dewey, Melvil.

1886 **Delisle, Leopold.**   Instructions pour la rédaction d'un inventaire des incunables conservés dans les bibliothèques publiques de France. (see Bulletin des bibliothèques & des archives. 1886. 3: 2-40.)

——— ——— 39 p. O.   Lille 1886.

116 specimen titles are appended. The general catalog of incunabula in French libraries for publication by the government is now being made by these rules.

1886 **Dziatzko, C;**   Instruction für die ordnung der titel im alphabetischen zettelkatalog der Königl. & universitäts-bibliothek zu Breslau.   11+74 p. O.   Ber. 1886.

——— Regole per il catalogo alfabetico a schede della reale biblioteca universitaria di Breslavia.   Prima versione dal tedesco, con aggiunte & correzioni dell' autore, a cura di Angelo Bruschi.   7+111 p. O.   Fir. 1887. (Biblioteca di bibliographia & paleografia.)

Reviewed in Centralblatt für bibliothekswesen. 1886. 3:289-295. Noted in L. j. 1886. 11:172 & 192.

1886 **Köhler, C;** Sylvio.   Die autorschaft & katalogisierung der akademischen dissertationen. (see Neuer anzeiger für bibliographie & bibliothek-wissenschaft. 1886. 47: 235-250.)

Criticized by Roquette in Centralblatt, 1887. 4: 335-342. For Köhler's reply & Roquette's answer, see p. 466-469.

1887 **Dziatzko, see 1886.**

1887 **Fumagalli, Giuseppe.**   Cataloghi di biblioteche & indici bibliografici.   19+199 p. O.   Fir. 1887.   (Biblioteca di bibliografia & paleografia.)

Reviewed in L. j. 12 : 547-48. Obtained prize offered by Italian minister of instruction for most rational rules for the preparation of catalogs & indexes. Its voluminous rules for carding follow Cutter closely. It treats briefly of alphabetical, classed & dictionary catalogs. Goes largely into history & general principles. Its bibliographies are full & valuable.   D: C. G.

1887 **Roquette, A.**   Zur frage der autorschaft älterer dissertationen. (see Centralblatt für bibliothekswesen. 1887. 4 : 335-342.)

Criticism of Köhler's article in Neuer anzeiger, 1886, 47: 225-250. For Köhler's reply & Roquette's answer, see Centralblatt. 1887. 4: 466-469.

1888 **Jewett, C: Coffin.**   Della compilazione dei cataloghi per biblioteche & del modo di pubblicarli per mezzo di titoli separati stereotipati, regole & esempi.   Prima versione dall' inglese a cura del G. Biagi...   9+120 p. O.   Fir. 1888. (Biblioteca di bibliografia & paleografia.)

1888 **Dewey, Melvil.**   Rules for author & classed catalogs as used in Columbia College Library with 52 fac-similes of sample cards; with bibliography of catalog rules, by M.. S. Cutler.   48 p. O.   B. 1888.

These are the A. L. A. Rules, as applied and enlarged in the Library School. They were in daily use by a large force of catalogers, and under revision for three years; we then printed and used two years, as a text book in the School. They are now brought out with many additions, illustrations, and clearer statement of points that caused difficulty to learners. It is intended to make these rules the best possible, and wherever improvement in statement or illustration is suggested or discovered, it will be incorporated in the next edition.

1889 ———...Library school card catalog rules; with 52 fac-similes of sample cards for author and classed catalogs, with bibliography of catalog rules by M.. S. Cutler; 2d ed... with marginal alternative rules.   48 p. O.   B. 1889.

This edition differs from that of 1888 in containing marginal explanations, amplifications and variations from the previous rules, for which there is good authority.

In preparation:—Cutter, C: A.   Rules for a printed dictionary catalog. New edition. See 1876 Cutter, C: A.

In preparation:—Cutter, C: A.   Regole per un catalogo a dizionario. Trans. by Guido Biagi.

The Library Bureau, 146 Franklin St., Boston, has become American publishers for this and all the Italian series now appearing as the Biblioteca di bibliografia and paleografia.

INDEX

Figures followed by letters refer to rules; S. C. to sample cards.

46

Index

47

Index

Library School

Accession Book Rules

With sample pages

BY

MELVIL DEWEY M. A.

Director New York State Library; President American Library Association

BOSTON
Library Bureau 146 Franklin St.
1896

A. L. A. Standard Accession-book.

First of all records to be filled, and by no means last in importance, is the book of accessions, the history of the growth of the library. To this the librarian turns for final reference in doubtful cases. Here is the complete story of each volume, fully told, but in the most compact form. It is the official indicator for the whole collection. Each line is a separate pigeonhole, in which, if not exactly the book, all the condenst facts about the book are placed. Thence they are never removed ; they are not stolen, or loaned, or condemned, or withdrawn, or sent to the binder, or lost. The card is never misplaced, the entry does not mysteriously disappear, a new edition never supersedes. Once written, "it is enough," til the paper grows thin with wear and the binding crumbles with age or the ink-lines entirely fade out of ken. He may turn to his book of accessions to learn *what*, and *where*, and *when*, and *whence*, and *how much*, and feel sure of his answer. A well-made accession-book has an element of mathematical exactness unknown to any other catalog. It is the *editio princeps*.

Every volume has a line, and the book is thus an indicator for the entire collection. By this complete, unchangeable record the additions for every day, week, month, and year are shown at a glance ; also the total number of volumes which the library has had ; and its present number by subtracting the total withdrawn and lost. This book is the most permanent of library records. There is no danger of losing or misplacing entries, as sometimes happens in card catalogs, nor of being compelled to rewrite them, as often happens in the shelf list.

The name catalog should not be applied to the accession-book or to the shelf list, but is restricted to the author, title, and subject catalogs, made primarily for the use of readers, while these are chiefly for official use. Commonly "accession" is used also as an activ verb. Some object to this, and prefer entry-book and "to enter ;" or record-book and "to record" or register and "to enter." "To register" confuses with the registration of readers. Additions or addition-book and "to add " is perhaps best, as entry on this record is technically the only way to "add " a book.

For this essential book many forms have been used, but the best features of all were finally combined in the A. L. A. Model, made by a committee of experts who compared thuroly all the various forms collected, made and tested samples, and finally agreed on all the details of materials, ruling, printing, and binding.

The double page is divided by double lines into three groups. First on the left, after the date of addition, are the three numbers assigned by each library, viz., accession, class or shelf, and book, followed by the volume number. Next come author, title, and imprint entries, which belong alike to every book of the edition regardless of any library. Lastly come the notes of binding, source, cost, changes, loss, re-binding, sale, etc, this third group like the first pertaining solely to this copy and library : or in brief, the line is filled with the title, preceded by the special library numbers, and followed by the special library notes.

The desirability of the location number in the accession-book has never been questioned, but the frequent changes in that number as ordinarily used made its use impracticable. The best managed libraries now assign permanent numbers to their books, so that they may be called for from the oldest edition of the catalog as readily as from the latest. With such a system it is a great convenience to refer directly to the shelf where the book may be found without consulting intermediate catalogs, also to glance down the column of numbers and see in what proportion the various departments, as indicated by those numbers, are receiving additions. The decision of the A. L. A. Committee was unanimous in favor of these columns, for it is believed that every library will sooner or later adopt a relativ location and so need them.

The old arrangement had the *volume column* with the other imprint entries on the second page. Here, the volume immediately precedes the author column, and at the first glance it is apparent what the entry is ; *e. g.*, v. 47, Harper's Magazine, is vastly more convenient than to follow across an entire page to the former place of the volume, with the attendant danger, both in entering and consulting, of getting on the line above or below, and thus making serious blunders. Practical use will convince those doubtful of the utility of the change. A still stronger reason is, that the volume-number is an essential part of the call-number or press-mark by which the book is found, and it is inconvenient to have the first part of this number at the beginning of the long line of entries, and the last part at the other end.

The colored lines help the eye in passing across the page, following one or two above or below, and thus avoiding the danger of getting off the line in crossing the fold.

4

It is economy to have a leather or canvas cover for the accession book while it is being filled. The cover is readily transferred to the next volume and lasts for many years, and as the volumes are filled they come out fresh and clean for preservation on the shelves, while without the cover they become very shabby and soiled in the course of necessary handling while entering 5,000 or 10,000 volumes.

It is cheaper to use the book with 10,000 lines for a library that expects to attain that number of volumes within any reasonable time; and for libraries of over 20,000 it pays to have a wood leger case with partitions making a pocket for each book, to preserve it safely and conveniently. A steel leger clip should also be used, to save needless handling in finding the place for current entries.

RULES FOR ENTERING.

In the following rules a choice of two or more ways is given in several cases. Each user should, before making the first entry, read these rules thru carefully and cancel all the forms mentioned, except those to be used, and add neatly in manuscript any added rules that seem desirable, and a " List of special abbreviations used in this book," which will show names of agents, funds, etc., local in character, but occurring so often as to need contraction.

This done, a glance at the preface will for all time show what the rule was for entering in that volume, and explain any abbreviations not, on the printed list.

1. *Enter each book immediately after it is collated and agreed with order-book and bill.*

a. The official business record of additions should be kept as strictly up to date as a cash account. If more books come in than can be written up at once, and some are wanted in special haste, they may be entered first, but under no circumstances should leave the library til properly added. Once bearing the accession-number, other facts are readily found; but a book without this guide is easily lost or confused with books from other sources or coming in on other dates. The accession-book corresponds to the invoice-book of a business house. As a package is opened, collate it with order-book and bill, to see that it is what was ordered, that the book is complete and in proper condition, and that the price is right. Check the price on bill, and enter it on order record, thus "agreeing" them. Then, if correct, enter the book *at once* on the invoice or accession-book. If incorrect or imperfect, do not enter it at all, for this fact decides that it is not to be "added" to the library

5

b. Pictures, statuary, maps, and all articles added to the library should be accessioned and numbered, so as to record the date, source, cost, and any other items of interest. Cards may also wisely be added under the names of artist and subject in the catalogs. It is well to have a special book of additions for these articles, if they are likely to be numerous; otherwise 100 or more numbers may be saved in a block at the front or back of the accession-book, thus keeping these peculiar articles together. To distinguish them and their numbers from books, prefix A, marking the first work of art A 1, and so on, the last number showing the extent of the collection. If wisht, a similar list for scientific specimens can be made under S 1, for maps under M 1, etc.

c. Maps, charts, etc., not in book form are stampt in the lower right corner, or near the title of the map. In the size column, length and bredth are given in cm ; e. g., 41 x 52 cm.

2. *Use all the standard library abbreviations in all entries.*

These include dates, authors' forenames, place, size, binding, etc. See Library *Abbreviations* appended to these rules for full lists. So many facts are given in so little space, that it is important to save room by using abbreviations ; and, as only those familiar with them use this book, the objection against their use in public catalogs does not hold good. All obvious contractions may be used in this book, specially in titles ; e. g., "geog." for geographical, "hist." for history, "biog.," etc., etc. The compact L. B. dates are best for all library uses.

3. *Give day, month, and year in the upper left margin of each left-hand page, and the day and month before the first entry of each day.*

The running date at the top is used in turning to find books by dates. It applies to all entries til a new date is prefixt to the accession number. If a whole page or more are received the same day, the date in top margin is enough. Sometimes only a single book is added, but its date must be given as carefully as for the pageful, for dates, like other figures, are almost worthless if not exact. This date shows when the book came into the library; and, if the accessions were written up daily, it would also be the date of entry. A large number coming at once should be entered under the same date, to show that they came together, even if the entry takes several days. It is less account what day the line is written, than what day the book was received into the library. Some however give the date of entry rather than reception when different, as being easiest. Others note both dates. This shows one reason for strict observance of Rule 1. If delay is necessary put the date on the cover where it will be covered by the book-plate, unless the order clerk puts, as he ought, the date on the inner margin of the first recto. If books accumulate they should be kept in order of reception ; and, if any are specially wanted before the others, the lines may be counted off so as to accession in proper order.

4. *Give to each volume the next consecutiv number on the first blank line of the accession-book, and enter this number on the lower margin of the first recto, after the title page of each volume received, and never assign the same number to another volume, even if the original be lost, sold, exchanged, or condemned, and an exact duplicate obtained.*

a. An accession-number is given each separate volume, and not to works, or sets, or lots, or series, or collections. Numbering *works*, in however many volumes they may chance to be, always leads to confusion. The last number should show how many volumes the library has received from the beginning. Books are often issued in parts and at intervals, like periodicals. If an effort is made to number books or works, rather than volumes, the first volume received in continuation makes trouble ; e. g., v. 4 comes in to-day and should be numbered 1347 ; but v. 1, 2, and 3 are numbered 975. That entry must be found and altered to show that v. 4 is here. When v. 5 comes in, it must be again altered, and so on to the end. Such constant changes and erasures make a slovenly book, and are sure to beget confusion.

b. To number each lot as it comes, without trying to keep all the volumes of a set together, is a little better, as it saves changes ; but, if an accession-book is kept at all, it is best to keep it properly, and let it show what it professes — the additions of each day in the exact order of their reception, without classification in any form.

c. In assigning *call-numbers* or *shelf-marks* by which readers call for what they want, books, and not volumes, are numbered. In the accession-book, *volumes*, not *books*, bear the number.

d. The rule gives a separate line to each volume. Some catalogers have put sets of 50 or 60 volumes all on one line. The only gain is a little paper ; for the apparent saving of labor proves no economy in the end. The entries, if the same, are *dittoed* with labor too trifling for mention. One serves for the whole line, instead of dittoing each word ; and even this is done only once in the life of the book. A single accession-book has 10,000 lines, or pigeon-holes for 10,000 distinct volumes. Trial of various plans proves it best to assign one of these pigeon-holes or lines across the book to each volume.

e. The rule assigns a given line to a given volume, and forbids its use for any other than that identical volume. There is no trouble then in recording titles, imprints, cost, source, binding, etc., for the different volumes of a set. If any volume is lost, or re-bound, or requires any note or comment to preserve its history and the record of its present state, the way is perfectly simple. If two or more volumes are put on a single line, confusion is sure to arise sooner or later, while by this rule any fact concerning any volume can be entered and found with the least possible labor.

f. Some libraries in replacing a lost book give it the same accession-number as the original. This is convenient and best for shelf-numbers, but is all wrong for *accession*-numbers. Lost books often come back after many years, and some day a wearisome effort to agree accounts discloses *two* books bearing the same accession-number. A book put in the library to-day in place of one lost five years ago was added *to-day*, and not at the time of the first purchase. It is e. g., the 1374th volume added to the library, and is to *take the place* of 975, which some one has lost.

g. Pamflets not previously accessioned, when bound, are entered the same as new books, on the date when they come in from the bindery, which is the time of their reception as books. In the source column the word "Binding," in place of the agent's name, shows that the pamflets had been in the library, but not entered.

h. In binding together two or more books already accessioned, the number of the first may be retained as the number of the collection, as it is the identical book, and all of it, with additions. In the Remarks column opposit each of the other volumes enter "Bound in No.——" with date.

i. This number as soon as assigned is stampt on the first recto after the title. When the book is re-bound, the number is preserved for immediate reference or identification after the book comes back from the binder. This is the most convenient place for this number, after the title itself, which it would deface, and the reverse of the title where the ink often shows thru or blots, and where it is harder to get a smooth surface for stamping. Even if given on the book-plate, it is repeated here as the plate is lost in re-binding.

j. If a numbering machine is used, stamp the accession number on the back of main author and subject card *bottom-side up*, so as to be read from the front when lockt in the catalog drawers, or crosswise of the back on the reverse of the upper left corner of the front. This number on the card often saves a double reference, and, the machine being set, it is no appreciable labor to give the number wherever wanted. On the back it takes no space available for title or notes, and is just as convenient, if stampt bottom-side up. When at the accession machine it is not known how many cards will be written, so only the one main author and one main subject card is stampt. This is also an advantage, in that a card not stampt on the back is thus recognized at a glance as an "added entry" or secondary card. The numbering machine, tho costly at first, is a real economy in a library, where it saves its cost in time of clerks and catalogers, beside giving compact printed numbers of the highest legibility.

k. Numbers when in column are quicker to write and easier to read if only the last two digits are given except for each 10th number. For 60c. a 1000 extra, the book can be had with the numbers all printed in advance. This is much the neatest and in the end the cheapest way.

l. Numbers cannot be printed in advance if dates of entry or receipt of books are given on a separate line. This is sometimes done because a blank line (with only the date in the center) above and below the list of books received together sets them off distinctly as a group by themselves. This plan, however, besides making the very desirable printed numbers impracticable, breaks up the uniformity of numbers beginning each page, and reference is less ready. By our rule of a line to a number, and dates in the left margin, every page begins with an even 20, and the eye is greatly helpt in quick reference.

5. *Enter the class, book and volume numbers as soon as assigned, in ink if permanent, in pencil if liable to alteration.*

a. Some libraries are so numbered, or liable to so frequent changes in their shelf-numbers or press-marks, that it is best to give only volume-numbers, which remain fixt. Others give the press-marks in pencil, so that they can be readily altered. The columns can be left blank if the system does not admit of their satisfactory use. They are of very great value to libraries that have a shelf-number not liable to frequent changes. 1. They refer directly to the shelves and shelf lists without consulting the catalogs ; 2. They make analysis and statistics vastly easier. They are the best check to show that all books paid for really get on the shelves, insted of disappearing by accident or design before getting on the inventory [shelf list] and catalogs.

b. If the old fixt system is used, alcove, range, and shelf, insted of class, will go in the first column, for which the heading CLASS was chosen, for its brevity and applicability to almost any system. Nearly all libraries agree in using a book-number between this and the volume-

9

number. Where the alfabetical arrangement is followed wholly or in part, these columns will be used for the words or letters and figures which determin the location of the book. These two columns are left blank til the book is cataloged, clast and shelf listed ; for it is impossible to give the numbers accurately, til the shelf list is consulted. Even in the alfabetical systems there is a chance that the identical combination has occurred before, and must be differentiated.

c. In the volume column, two volumes bound in one would be entered 1 & 2, 3 & 4, etc. One volume bound in two parts would be entered 1.¹, 1.², etc., each part on a separate line.

d. Give the number of the volume if more than one. If in only one volume, leave this column blank as 1 indicates that it is the first of a set of two or more. Make all entries of facts perfectly definit.

6. *Give the author's name and the title, as in the brief-title finding index. (For detailed rules see Library Notes, v 1., p. 111-131.)*

a. Space allows only a brief title, and other facts are given with so much fulness that the book is readily identified. If the work is anonymous, leave the AUTHOR column blank, to be filled when authorship is discovered. The line separating author and title is faint, so as to be seen only when lookt for to guide in making the titles line accurately one under the other. When author's name or title is very long, this line is simply disregarded, but in most entries there will be a little space between the author and the beginning of the title.

b. The form of author's name to be used, or the heading which takes its place in the author column for transactions and various other books, must correspond with the entry in the other catalogs ; and, as it must be decided from them, a careless heading should not be written here at the risk of error. With this as with the class number, if in doubt, and the book must be entered at once, leave the author column blank, and fill in after the heading is decided for the other catalogs. A volume of pamflets is entered under the heading used on the main card with a note ; e. g., "& 9 other pam."

7. *Give the place, publisher, date, pages, and size, in accordance with rules for full titles, except that more abbreviations may be safely used.*

a. If several places or publishers are printed, give only the first named on the title ; or the most important if the main publisher is given in large type with fine type names preceding. Some omit the publisher, but it should be once recorded somewhere to help identify each book or find a duplicate if needed. The publisher differentiates two editions of same place and date, and is often used in replacing, tracing, etc. If omitted from all other records, give it here. Leave space between abbreviation for place and publisher, so it shall not look like a name and initials ; *e. g.,* L.   MacMillan, *not* L. MacMillan.

b. If something must be omitted to save labor, perhaps the publisher can be spared easiest, or the pages, if given on the cards.

c. Give the date of publication in years of the common calendar, and in Arabic figures.   Never, here or elsewhere, use Roman numbers.

⌄ d. If it differs more than a year from publication date, always give year of copyright with *c* prefixt.   The line headed DATE is wide enough for six figures, so this important item specifying the real date of publication can be added.

e. The extent of a book depends on pages and size, so that both should be given.   The paging is the most useful item that can be given in so small space ; as it is on the cards it can be filled in with the class and book numbers without looking it up twice.   It is one of the most useful items in identifying similar editions.   A pamphlet is only a thin book, and is best described by giving paging ; e. g., "37 p. O."   All the sizes (binding, paper, and type), as well as fold, in case it should be desirable, can be given by interlining, but except in rare books, the size-letter is sufficient.   Give the size by the A. L. A. book-size rule.   If it is on the line, call it the larger if it has been trimmed in re-binding ; if untrimmed call it the smaller.   For books paged in fragments, do not substract, but give the first and last ; e. g., p. 613-1120.

f. An admirable help in quick and accurate sizing is Cole's Size Card, publisht by the Library Bureau.   It gives by diagonal lines the proper prefix *nar, sq,* or *ob,* without separate measurement.   It is kept inside the cover ; some mark the size lines inside the front cover of the accession-book or paste in a Cole card, tho it is less handy to lay the book accurately on the card than to insert the card under the cover, which serves as an automatic guide to hold it in exact position.

g. The size of type is rarely given, tho to many an important item in choice of editions for reading.   It is a part of size or extent, but is more useful on the catalogs than here.

ˋ 8. *Give the binding material, indicating half binding by prefixing ½, or better a superior [2]; e. g., [2]mor. ˙If the book has leather corners also, ¾, or a superior [3] will indicate it.ᴀ If by a famous binder, or otherwise remarkable, note in Remarks.*

9. *Under* SOURCE *write the name of the giver, if a gift ; the name of the fund if bo't from the income of a special fund ; or the name of the firm or library agents, of whom bo't, if from the general fund.*

11

Some give the name of the supplying agents in all cases, prefixing the initials of the fund in the second case. The funds of each library are so well known that the initials are ample, so the column for SOURCE allows room for both agent and fund. Enter at least the initials of the agent in all cases.

10. *Under* COST *give in dollars and cents the actual cost of the book, including exchange on foreign books.*

a. A ruling for pounds, shillings and pence is not needed. So few books among the mass in the library will be so billed, that it is waste of space to devote three whole columns to these headings. Even in these cases convenience requires that cost be given in ordinary denominations, so that a moment tells an inquirer the cost of any book. If the exact amount in foreign money must be kept, interline it in the cost column, or better, to allow footing the cost for statistics, enter it in *Remarks*, next to cost column. If list price is given, it is useless to repeat the net price also in foreign terms.

b. If the list price is recorded, put it at the right of the source column, next to the left ruling of the cost column, using fr, m, and s for francs, marks, or shillings, and unmarkt figures for dollars. If, because of rarity or *net* list price or special terms from auction or second-hand dealers, the price paid is *more* or *less* than usual for a book of that list price, prefix "n" (net) to the cost given, to show it is not an error in entry. Some wisely do not stop to look up list prices, but enter them only when already known.

c. When several volumes are bo't at once, give cost of the series opposit the first entered, followed by a note of the number of volumes included; e. g., v. 4, 5, and 6 of some work come in together, and cost together $13.44. Instead of dividing this up, and entering $4.48 against each volume, make the entry against the first (v. 4) in this way : 3 v. $13.44. Or (a more difficult entry to make neatly), connect the lines of the different volumes by a bracket, and write the cost against the center. For convenience in identifying what came on each bill, the date and total may be given in *Remarks* opposit the first entry ; e. g., on line 10,431 the remark for a bill covering 30 v. would be "To 10,460 is bill of 19 N. $110.⁶²." The date of the bill is always earlier than that of reception. 10,460 shows the line of the last entry included in the bill, as does the remark of the next bill, unless it is for a single volume, and so not recorded. It is often handy to check off special bills in this way, and the labor is trifling to thus indicate the extent of large bills. Mark gifts "g" in cost column, and if cost is known add it in [ ]; e. g., g [1.75]. For convenience in footing amount paid out for books and amount of gifts, it is well to enter the cost value assigned to gifts in red ink, as being more distinct than the [ ], and showing more grafically the proportion of gifts to each page. Some even write the entire line in red. Give items of cost carefully, thus making the accession-book for all practical purposes the invoice-book.

11. *Mark the cost of each book in inner margin of first recto after date of reception, when collating with the bill.*

From this place it is copied in the accession-book, and is often found of great convenience in determining value without consulting records or bills. The cost written in the inner corner of some special page agreed on in each library, serves to identify books with labels removed, or covers taken off in binding, or by accident, or by design where theft is intended; but any reader may wish to know the cost, and it is unwise to use it as a detectiv mark, of which the place cannot be told to all interested. A much safer private mark is a perforation with an awl or pin in a secret place; e. g., a pin hole thru the center of the fifth o used in the paging. This can hardly be found by accident or removed, if known, so as not to be detected by an expert.

12. *Under* REMARKS *indicate any re-binding, sale, loss, exchange, withdrawal as duplicate, binding in with another volume, or any change or disposition.*

The preceding entries tell what the book was when it came into the library. REMARKS tell of any changes, and of the final disposition in case the book is no longer in its accustomed place. This rule requires less labor than at first appears, and saves more than it costs. When books come in from the bindery, it is a very brief matter to open to their numbers, and note the new dress, with its cost. Then if the volume be lost and the reader wishes to pay for it, there is a means of knowing whether it was in paper as at first bo't for 25 cents, or in half morocco as re-bound, at an added cost of $1. The accession-book is the book of final reference for these technical facts, that appear on no other catalog. The efficient librarian must be able *somewhere* to refer to everything of the kind, and for this no other record offers so great advantages.

The librarian who will keep an accession-book on the plan above described, will find himself well repaid It will be in constant requisition, the final authority to which will be referred all doubtful questions regarding the past history or present state of any one or of all his books.

Date	Number	Bind'g	Source	Cost	Class	Book	Vol.	Remarks
			Since 1 Ja 86	842 20			cop 1	
	60101	nor.	U.S.	g³ 94	027.073	Un3	1	
	02	sap.	Bgs Carey fund	1	"	"	cop 1 2	bd. ² mor. 20 Jl 88
	03	l	W.H. Shepard	3 25 60	014.2	T36		
	04	sap.	Putnam	34	973.71	B39		bd. ³ ro. 1 Jl 88 60121
6 Je	05	l "		2	973.74	D66		lost 15 Jl 88, repl by ^ 1089
	06	" "		84 75	813.43	Q		cop 6, withdrawn, wornout
	07	ro.	Binding		304	C18		75c.
	08	cf.	ex. Y.M.A.	1	883.1	Q		cop 2 sold 8 S 88 to W.B., dup. 20 D 88. ex. J.
	09	l.	Putnam	84	813.49	M94		Same 22 D 88 60c.
	60110	rel	Bangs	25	230	Es6	1 & 2	
	11	l	Stechert	1 50	913.38	M27		dup 16 N 88
7 Je	12	sap.	J.H. univ.	25 55				bd. in 60124
8 Je	13	ro.	Ames lib.	g4	019.1	Am3	pt.1	rpd. 10 89
	14	" "		2 v. "	"	"	" 2	
	15	f.	St. Education fund	4	378.42	H86	2¹	
	16	"	"		"	"	2²	
	17	ds.	Bangs	06	385	D19		
	18	ch.	W.S. Bissell	g 25	511.8	I3		
	19	us	U.S.	g6 ?	557 87	N6		
	60120	ds.	St. N. soc.	g 25	367	Sa2		
	21	l			973.74	D66	cop.2	replacing 60105
	22	ap.	Bgs Carey fund	2 v. 2 2	027.073	Un3	1	bd. ² mor. 1 Ja 89
	23	"	"		"	"	cop.2 2	"
	24	nor.	J.H. univ.	3 94	305	J62	1	bd. ³ mor. 17 O 89
	25	f	ex. G.H. Page	2 25	320.1	L62	2	
			carried forward	870 56				

No.	Author	Title		Place	Publisher	Date/notes		Call no.	Notes
101	U.S.-Education bur.	Public libraries in U.S.		W.	Govt	1876 35,1187 O	mor. U.S.	027.013 Un3	2 Ad. 'mor 20 X 83
02	"	"		"	"	59	pap. Byo Carey fund		
03	Thomas, R.,	Handbook of fictitious names		S.	Smith	1868 14,1,234	cl. W. H. Shepard	3.25 014.2 T36	
04	Bucher, H: W	Speeches on the rebellion		N.Y.	Lovell	1857 368 D	pap Putnam	34 913.71 B39	bd. 'no 1 Jl 88
05	Dodge, J. O.	Bird's-eye view of our civil war		B.	Osgood	1883 11-346 O cl	.	2. 913.74 D66	last 13 X 89, repl by
06	Howells, W: D.	Chance acquaintance		"		1877,2,279 S	. .	84 813.43 Q	copi withdrawn, worn out
07	Carey, H: C.	Credit system in France, Gt Br & U.S. & other pam.		Ph.	Carey	1838 130 O	no Binding	304 C18	
08	Homerus	Iliad Buckley, tr.		S.	Bohn	1851,2,466 D	cl in Y.M.Q.	883.1 Q	copi sold 8 S 86 6 Jl 12.
09	Murfree, M.N.	Where the battle was fought		B.	Osgood	1884 423	cl Putnam	84 813.49 M94	
10	Espagne, Jr de	Alle de wercken		Utrecht	Ribbius	1678	S	25 230 Es6	152
11	Mahaffy, J. P.	Social life in Greece Ed 2 enl.		S.	Macmillan	1875 15,416 D	cl Stechert	150 913.38 M27	dup 1611 88
12	Adams, H. B.	Norman constables in Amer.		Balt.	J.H. univ.	1883 38 O	pap J H univ.	25	bd in 60124
13	Ames free lib.	Catalogue		B.	Rand	9+442 Q	no Ames lib.	g 019.1 Am3	pt I apd 1083
14	"	"				443-883			2
15	Huber, V. A.	English universities		S.	Pickering	1843 16,2,418 O	cl St Education fund	4 378.42 H86	2
16	"	"				Iv 413,740			2
17	Dana, W: B so pub.	Value of railroad securities		N.Y.	Dana	1882 ump	dT bds Bangs	06 385 D19	
18	Hoppus, S	Tables for measuring Ed 18		S.	Rivington	1823 60+226 naO	cl U.S. Bussell	g 25 511 8 13	
19	U.S.-Geog's geol. survey	Atlas, for, Report on geology of Uinta mts.		W.		1876 2 mapfol F	bus U.S	gd 557 87 N6	
20	St Nicholas soc. of N.Y.	Record of the semi-centennial anniversary		N.Y.	St. N. soc.	1885 42+2, O	bds St. N. soc	g 25 367 Sa2	
21	Dodge, J. O.	Bird's-eye view of our civil war		B.	Osgood	1883 11,346	cl	913.74 D66	replacing 60103
22	U.S.-Education bur.	Public libraries in U.S.		W.	Govt.	1876 35,1187	pap. Byo Carey fund	027.073 Un3	bd 'mor 1 Ja 89
23	"	"		"		59			
24		Johns Hopkins studies in hist. & polit. sci.		Balt.	J.H. univ	1883 v p	mor J H univ.	3 305 J62	bd 'mor 17 D 88
25	Sieber, F.	Manual of polit ethics		B.	Little	1838 16+668	cl in G.H. Hoye carried forward 81836	225 320.1 L62	2

☞ *For full list of Library Abbreviations see Card Catalog Rules,*
p. 35–40.

☞ *For full list of Library Abbreviations see Card Catalog Rules,* p. 35–40.

Library School

Shelf List Rules

With sample pages

BY

MELVIL DEWEY M. A.

Director New York State Library; President American Library Association

BOSTON

Library Bureau 146 Franklin St.

1898

Shelf list rules

These rules have been worked out by long experience in the use of shelf sheets 25 x 20 cm, the size in common use among libraries. Variations necessary for a sheet 10 x 25 cm are specified. A sample page of each size is appended.

ARRANGEMENT OF ENTRIES. Arrange entries as books are arranged on the shelves — first, by class number, treated decimally; second, by book number, arranged (a) alphabetically by the capital letter (b) numerically by numbers following the capital, treated decimally.

NUMBER OF SUBJECTS ON PAGE. In beginning a new shelf list, or a new sheet, never enter more than two subjects on a page, (or one subject on a sheet if 10 x 25 cm sheets are used), thus allowing for intercalation, but leave no lines between book entries.

LARGE SUBJECTS. Begin *all divisions* like 110, 220, 370, 450, and such sections as 511, 512, etc., also in Literature, 811.49, 813.49, etc., and any subject on which you are likely to have a large number of books, on a new sheet, and leave the rest of the sheet blank, beginning the next subject on a fresh sheet.

PERIODICALS. In case of periodicals and transactions of societies like 052, 305, 906, etc., put not more than one entries on a page of periodicals still being published. Leave room for accession numbers of volumes to be added, arranging these in three columns, thus : —

36723-4	v. 1–2	50002	v. 14	69467	v. 26
44321	v. 3	68876	v. 15	73125	v. 27

If 10 x 25 cm sheets are used, put one entry on a sheet, and arrange accession numbers in four columns.

CLASS NUMBER. Write class number in blue ink on upper line after "class," and, if a second subject occurs on the same page, put its class number on outside margin of sheet on the line above entries to which it belongs.

BOOK NUMBER. Write book number in its column in black ink, and in heavier figures to make it more prominent.

ACCESSION NUMBER. Write accession number in its column in black ink, with – between inclusive figures ; e. g. 3342-7.

SIZE. If a book is over O size, put its size, Q, F, etc., in red ink in upper right corner of space belonging to accession number. If size is incorporated in the book number, no other size designation is needed.

3

VOLUMES. Put number of volumes, if more than one, in "vol." column in black ink; if the library contains part of a set, write; e. g. v. 1, *or*, v. 2–6; *write also* 2 in 1; 6 in 3; v. 9¹, v. 9². In case of annuals where there is no volume number, use the year in volume column. On cards, Special loc. letter goes before

SPECIAL LOCATION. When necessary to indicate location of books removed from their regular place, write in blue ink in upper right corner of book number column "Ref.," or number of room, or other needed designation. ' or stamp

INDEX VOLUMES. Place an index volume if unnumbered at the beginning of a set, instead of intercalating after the last volume covered by the index, and mark it with volume number 0. If more than one, mark them thus, 0¹, 0², 0³. Add in ~~title~~ Volume column, years or volumes covered; e. g. Index, 1840–50.

SEVERAL COPIES. When there is more than one copy of a book, give a separate line to each, indicating the number of the copy in red ink in volume column; e. g. 49725   cop. 1

49772   cop. 2, unless the accession numbers are consecutive; e. g. 33968–69 2 cop. If the work has more than one volume, write number of copy in volume column above volume number.

AUTHOR. Write author's surname in black ink in its column, using no punctuation. Do not give forename except to distinguish from another author of same name in same class number. In case of two joint authors write surname of each; if more than two, write surname of first and "& others."

TITLE. Write brief title in its column in black ink with no punctuation. This title can be shortened from title page, or taken from running title, or from back of book, if the book is known by that title. If the shelf list is to be used chiefly as a stock-taking book, use binder's title. If used mainly as a subject catalog, use any well-known title or a shortening of the title page, taking great care in abbreviating title that it may be at once clear, short, and comprehensive. Add in briefest form the name of series if well known, (e. g. Am. men of let.)

EDITIONS. In case of two editions of the same book, add date of each, or number or name of edition, or editor's surname after title. In the classics add editor's surname.

NEW ENTRIES UNALPHABETED. In adding to a shelf list leave one line vacant after the continuous alphabet, and add new entries as they come, with no attempt at alphabeting. As often as expedient rewrite the sheet and combine into one alphabet.

DATES. Put on upper margin of each sheet the date when it is

4

started; e. g. 13 Je 83. When withdrawn add after a dash the date of rewriting; e. g. 13 Je 83–17 Ap 90.

OLD SHEETS. Arrange withdrawn shelf sheets in order of class numbers and save for reference.

CHANGED NUMBERS. If the class number or any part of it is changed after the book is shelf-listed, do not erase entry, but draw a red line through it, and write in red ink in author column the new class and book number, and enter like a new book in its new place, changing the numbers on book, accession book, and all catalog entries. If book number alone is changed, draw red line through it, and write in red ink the new number above; and if the entry is in an alphabet of more than 10 items, cancel the whole entry by a red line and reënter as if a new book; this helps in tracing the book. When the sheet is rewritten, leave it out.

www.ingramcontent.com/pod-product-compliance
Lightning Source LLC
Chambersburg PA
CBHW020335090426
42735CB00009B/1550